<u>Disclaimer</u>

Book Title: Best Practices Guide for High-Volume Fly Ash Concretes: Assuring Properties and Performance

Book Author: Dale P. Bentz; Chiara F. Ferraris; Kenneth A. Snyder;

Book Abstract: A best practices guide is developed from a synthesis of recent research on high-volume fly ash (HVFA) concrete mixtures. These best practices can be applied by the concrete construction industry to achieve desired properties and ensure the (high) performance of HVFA concrete mixtures in practice. As such, the report considers all aspects of HVFA concrete production, from the characterization of the starting materials, through mixture proportioning and curing options to achieve desired properties, to the in-place early-age and long term performance of the concrete in its fresh and hardened states. Both mechanical and transport properties are considered in detail. Perspective is established based on a brief review of current practices being employed nationally. Each topical section is concluded with a practice-based set of take away messages for the design and construction community. The report is intended to serve as a valuable resource to these communities, providing both a research summary and a guide to practical steps that can be taken to achieve the optimum performance of these sustainable concrete mixtures.

Citation: NIST TN - 1812

Keywords: Characterization; durability; high-volume fly ash; incompatibilities; setting; strength; ternary blends

NIST Technical Note 1812

Best Practices Guide for High-Volume Fly Ash Concretes:

Assuring Properties and Performance

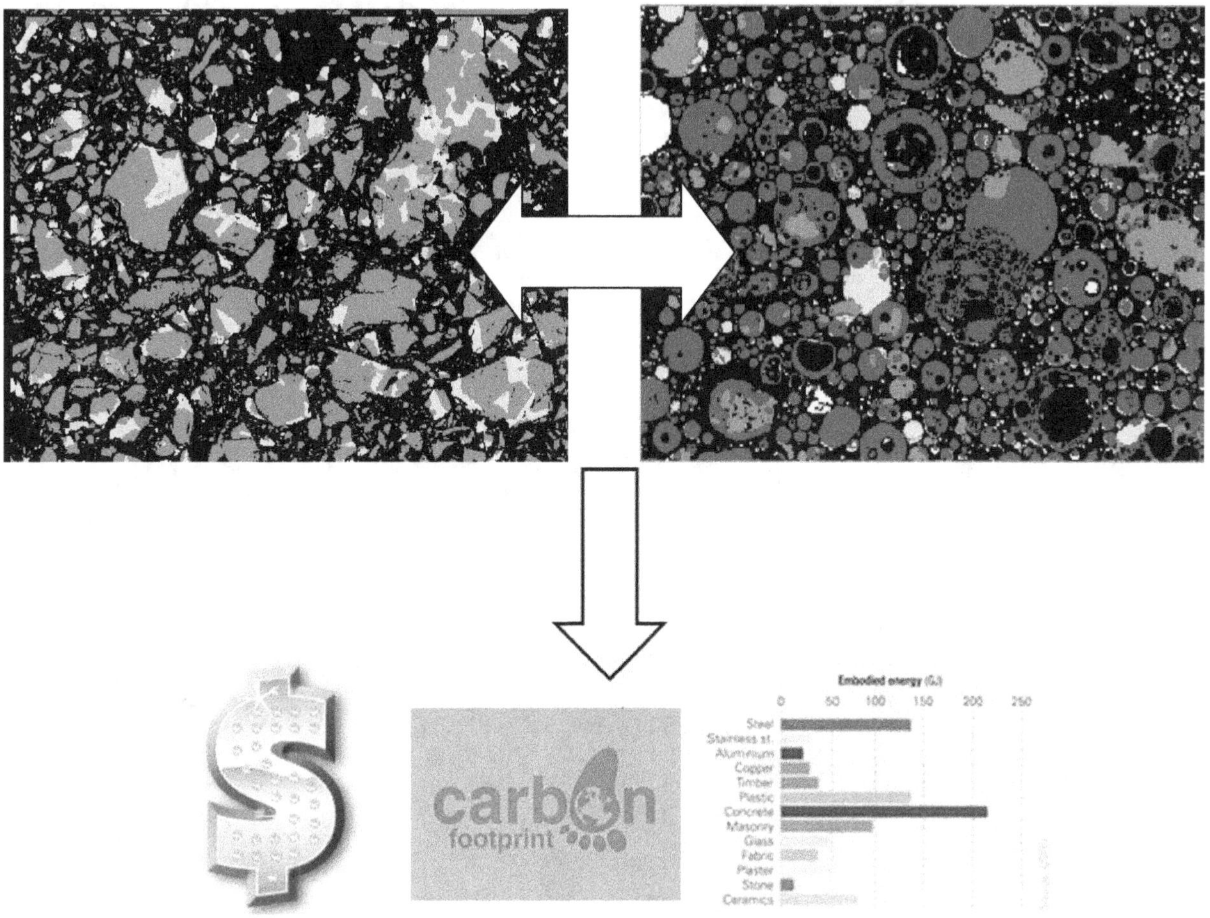

Dale P. Bentz
Chiara F. Ferraris
Kenneth A Snyder

NIST

National Institute of
Standards and Technology
U.S. Department of Commerce

Best Practices Guide for High-Volume Fly Ash Concretes:
Assuring Properties and Performance

Dale P. Bentz
Chiara F. Ferraris
Kenneth A. Snyder
Materials and Structural Systems Division
Engineering Laboratory

September 2013

U.S. Department of Commerce
Penny Pritzker, Secretary

National Institute of Standards and Technology
Patrick D. Gallagher, Under Secretary of Commerce for Standards and Technology and Director

National Institute of Standards and Technology Technical Note 1812
Natl. Inst. Stand. Technol. Tech. Note 1812, 56 pages (September 2013)
CODEN: NTNOEF

Abstract

A best practices guide is developed from a synthesis of recent research on high-volume fly ash (HVFA) concrete mixtures. These best practices can be applied by the concrete construction industry to achieve desired properties and to ensure the (high) performance of HVFA concrete mixtures in practice. As such, the report considers all aspects of HVFA concrete production, from the characterization of the starting materials, through mixture proportioning and curing options to achieve desired properties, to the in-place early-age and long-term performance of the concrete in its fresh and hardened states. Both mechanical and transport properties are considered in detail. Perspective is established based on a brief review of current practices being employed nationally. Each topical section is concluded with a practice-based set of recommendations for the design and construction community. The report is intended to serve as a valuable resource to these communities, providing both a research summary and a guide to practical steps that can be taken to achieve the optimum performance of these sustainable concrete mixtures.

Keywords: Characterization; durability; high-volume fly ash; incompatibilities; setting; strength; ternary blends.

Table of Contents

List of Figures

List of Tables

Introduction

The sustainability movement of the 21st century has provided impetus to various industry sectors to identify alternatives to their energy intensive (i.e., costly) materials and processes. In the concrete construction industry, a majority of the embodied energy and cost of concrete is due to the portland cement. Although the industry has been using industrial by-product materials for decades to replace a portion of the portland cement to achieve specific desirable engineering properties, recent efforts to significantly reduce the embodied energy have produced renewed focus on far greater replacement percentages of portland cement in concrete mixtures. One approach that has received considerable attention in recent years is the production of so-called *high-volume fly ash* (HVFA) concrete mixtures [1], where fly ash replaces cement at proportions well above the 15 % to 25 % commonly encountered in current blended cement concrete mixtures. As a reference point, according to a 2012 survey [2], fly ash usage in 2011 was estimated at 15 % of the total cementitious binder (portland cement, fly ash, and slag) in ready-mixed concrete in the U.S., with the usage consistently trending upwards during the past ten years. With much fly ash still being diverted to landfills and existing landfills filled with fly ash that could potentially be used in concrete, there is still an important opportunity to further increase fly ash contents in concretes. The aforementioned survey conducted by Obla et al. [2] identified the two primary causes preventing increased use of fly ash in ready-mixed concrete: 1) performance issues including setting time and strength gain, and 2) specifications that restrict usage. By providing information on best practices for the production, construction, and evaluation of HVFA concretes, this report will address the most commonly identified performance issues, paving the way for increased specification of this sustainable construction material.

A Brief Look at Current Practice

Numerous ready-mix companies are actively pursuing the development and implementation of HVFA concrete mixtures. The following summary is a composite based on interviews with two such producers [3,4]. HVFA concrete mixture proportions will vary widely depending on locality, as governed by materials' availability, shipping costs, and structural applications. Many specifications from state Departments of Transportation and elsewhere limit fly ash replacement levels to 25 % or less, on a mass basis.[1] Conversely, for projects seeking LEED certification, such as school buildings, 50:50 and even 60:40 (fly ash: cement) mixtures have been developed and implemented without significant difficulties. In some cases, the setting time delays typical of HVFA mixtures can be used to advantage, as they permit longer hauling times and may allow for a reduction in the dosage of retarders (chemical admixtures) used in the mixtures. The increased retention of workability (i.e., ease of mixing, placement, consolidation, and finishing) is another characteristic of HVFA concrete mixtures that is often used to advantage.

For flowable fill and low density applications, HVFA mixtures with up to 90 % fly ash have been developed. It should be noted that for particularly reactive fly ashes, it is possible to produce acceptable concretes with 100 % fly ash, as documented by Cross et al. [5] and as employed in certain commercially-available products. These ultra-high-volume fly ash mixtures provide enhanced pumpability and also improved thermal properties (in combination with the incorporation of low density fillers such as lightweight synthetic particles) for the preparation of thermally resistant grouts and concretes.

The U.S. Naval Facilities Engineering Command (NAVFAC) has recently developed an HVFA concrete formulation in which 50 % of the portland cement is replaced with fly ash on a mass basis [6], providing "constructability, cost-effectiveness, and environmental benefits." Specifications for this concrete included a minimum 28 d compressive strength of 5000 psi (34.5 MPa), a maximum 28 d drying shrinkage strain of 0.05 %, a maximum water-to-cementitious materials ratio (w/cm) of 0.4, suitable slump and finishability characteristics, the ability to place the concrete during all seasons, and a sufficient early-age strength to permit formwork to be removed at an age of 7 d. Mixture proportions for the existing control (lower fly ash content and some silica fume) and HVFA concretes for cold and warm weather are provided in Table 1 [6]. The warm weather HVFA mixture was modified from its cold weather counterpart by removing the accelerating admixture and reducing the cementitious content to take advantage of the enhanced strength development at warmer temperatures. The concretes were used to construct test beams, from which extracted cores were analyzed with respect to strength, absorption, diffusion coefficients, and drying rates. The cold weather and warm

---

[1] It should be recognized that the replacement of cement by fly ash can be performed on either a mass or a volumetric basis. Because fly ash is typically less dense than cement (e.g., about 2500 kg/m³ vs. about 3150 kg/m³), replacing a mass of cement with an equal mass of fly ash will alter the volumetric water content, cementitious powder volume fraction, unit weight, and/or yield of the concrete mixture. Replacing a volume of cement with an equivalent volume of fly ash allows one to maintain these quantities at their non-fly ash mixture values. Although it is not common industry practice, the authors of this report recommend that replacements of cement with fly ash be performed on a volumetric basis. This volumetric-based proportioning and replacement requires that the densities of the cement and fly ash first be measured, using the ASTM C188-09 Standard Test Method for Density of Hydraulic Cement or some other acceptable method.

weather HVFA concrete mixtures were found to meet the specification requirements, although the warm weather HVFA concrete mixture required 56 d for its lab-cured cylinders to reach the specified 28-d compressive strength (5000 psi), with the 56-d strength of the field-cured cylinders of this mixture reaching 4820 psi (33.2 MPa). The transport properties (e.g., resistivity) of the HVFA concrete mixtures were superior to those of the control mixture, with the diffusion coefficients and rapid chloride penetrability test (RCPT) values of the two HVFA mixtures improving significantly with extended curing from 90 d to 180 d. The study demonstrated that HVFA concretes with acceptable performance could be formulated to meet the original project specifications while also providing a considerable cost savings. The final recommendation from this report [6] was that the Navy should encourage the use of such HVFA concrete mixtures in all marine structures and in other structures commonly subject to corrosion of the steel reinforcement. The Naval facilities study [6] was conducted using a specific portland cement and a specific fly ash. Because the properties of fly ash can vary among sources, there is no guarantee that following the proportions in Table 1 will achieve the same properties and performance. The degree to which the concrete mixture will require modification, or whether the specified properties can be achieved at all, will depend upon the specific characteristics of the fly ash being employed. The mixture proportions in Table 1 are thus only provided to exemplify HVFA concrete mixtures that have worked well in practice and not to suggest definitive (generic) mixture proportions that could be employed universally.

Table 1. Mixture Proportions for Naval Facilities Test Concretes [6]

	Control	HVFA (cold weather)	HVFA (warm weather)
Total binder content[1]	860 lb/yd^3	660 lb/yd^3	564 lb/yd^3
Cement	660 lb/yd^3	330 lb/yd^3	282 lb/yd^3
Fly Ash	135 lb/yd^3	330 lb/yd^3	282 lb/yd^3
Silica fume	65 lb/yd^3	---	---
Water	266 lb/yd^3	199 lb/yd^3	169 lb/yd^3
Sand, ssd[2]	1040 lb/yd^3	1460 lb/yd^3	1468 lb/yd^3
¾" coarse aggregate	1607 lb/yd^3	1715 lb/yd^3	1782 lb/yd^3
High range water reducer (HRWR)[3]	60 fl oz/yd^3	66 fl oz/yd^3	56.4 fl oz/yd^3
Corrosion inhibitor	128 fl oz/yd^3	---	---
Non-chloride accelerator		132 fl oz/yd^3	---
Target air content	5 %	5 %	5 %
Water/binder ratio	0.31	0.30	0.30
Fly ash (% of binder)	---	50	50

[1] 1 lb/yd^3=0.593 kg/m^3
[2] ssd = saturated surface dry condition
[3] 1 fl oz/yd^3=38.7 mL/m^3

Another recently developed useful tool for concrete designers and specifiers is the Mix Optimization Catalog for Proportioning Fly Ash as Cementitious Materials in Airfield Pavement Concrete Mixtures [7]. This online, interactive tool provides guidelines and recommendations on mixture proportioning, fly ash properties, admixture selection, curing, and relevant standard tests for producing fly ash concretes for a user-selected pavement project. Projects are characterized by deicer exposure, aggregate reactivity, cement type (alkali level), opening time requirements, and paving weather. Fly ash recommendations are classified by calcium oxide content, fineness, loss on ignition, and replacement level. Fly ash replacement levels are classified as low (< 15 %), moderate (15 % to 30 %), high (30 % to 50 %), and very high (> 50 %) on a mass basis.[1] In addition to recommending a fly ash and a replacement level, the software also provides recommendations for admixtures and curing, as well as suggesting which ASTM standard test methods can be utilized to ensure quality.

Characterization

In the United States, fly ash is typically classified as either Class C or Class F based on its chemical composition, as defined in the ASTM C618-12a Standard Specification for Coal Fly Ash and Raw or Calcined Natural Pozzolan for Use in Concrete [8]. The major delimiter for this classification is the sum of the silicon, aluminum, and iron oxide percentages in the fly ash, being a minimum of 70 % for a Class F designation and a minimum of 50 % for a Class C classification. Other chemical requirements specified in the ASTM C618 specification (for both Class C and Class F fly ash) include sulfur trioxide content (5.0 % maximum), moisture content (3.0 % maximum), and loss on ignition (6.0 % maximum).

The ASTM C618 specification also details a series of physical requirements for the fly ashes, which are evaluated according to the ASTM C311-11b Standard Test Methods for Sampling and Testing Fly Ash or Natural Pozzolans for Use in Portland-Cement Concrete [9]. These include fineness, strength activity index at 7 d and 28 d, water requirement, soundness, and uniformity in density and fineness. Currently, the strength activity index test is evaluated by preparing a cement/fly ash mortar (mass-based replacement of 20 % of an ASTM C150 portland cement by fly ash) that has a variable water content, but equivalent flow to that of a control ordinary portland cement (OPC) mortar prepared with a fixed water-to-cement ratio, $w/c=0.484$, by mass. The measured strengths of the prepared cement/fly ash mortar at 7 d and 28 d are then divided by those of the $w/c=0.484$ control OPC mortar to obtain the strength activity indices, reported as a percentage. Because the densities of fly ash and cement are significantly different, this mass-based replacement procedure reduces the water and sand contents of the tested fly ash mortar relative to those present in the control. Furthermore, preparing the cement/fly ash mortar at equivalent flow can lead to an additional increase or decrease in the unit water content. As strength is strongly dependent on the volumetric (unit) water content of the mortar mixture, the relative effect of the fly ash on the strength activity is confounded due to any changes in water content. For this reason, an alternative procedure based on preparing the cement/fly ash mortar with identical volumetric proportions (sand, water, cementitious powders) to that of the control OPC mortar has been proposed and supported by research results [10].

It is generally recognized that the ASTM C618 classification is insufficient to completely characterize a fly ash in such a way as to anticipate its performance in concrete. In Canada, fly ashes are additionally classified based on their CaO content into three classes, as opposed to two. However, both of these bulk chemical classification schemes often fall short of providing the end-user with sufficient information to predict the reactivity and performance of a particular fly ash. It is generally the composition and proportions of the various glassy phases within a fly ash that influence its reactivity in a cement-based system, and these parameters are neither identified nor quantified in the current ASTM C618 specification. While crystalline phases in fly ash can be readily assessed using X-ray diffraction [11], the glassy phases are more difficult to characterize because particles can have variable chemical compositions. Valuable information on the chemical composition, morphology, and spatial statistics of both glassy and crystalline phases can be obtained by employing scanning electron microscopy (SEM) combined with X-ray microanalysis and multispectral image analysis [12-15]. As just one example, Figure 1 shows a processed multispectral analysis of a Class F fly ash in which eight different crystalline and glassy phases have been identified [14], including three different calcium aluminosilicate glasses. The all-important reactivity of these different (glassy) phases can then be analyzed by reacting these fly ash particles in synthetic pore solution, for example [15].

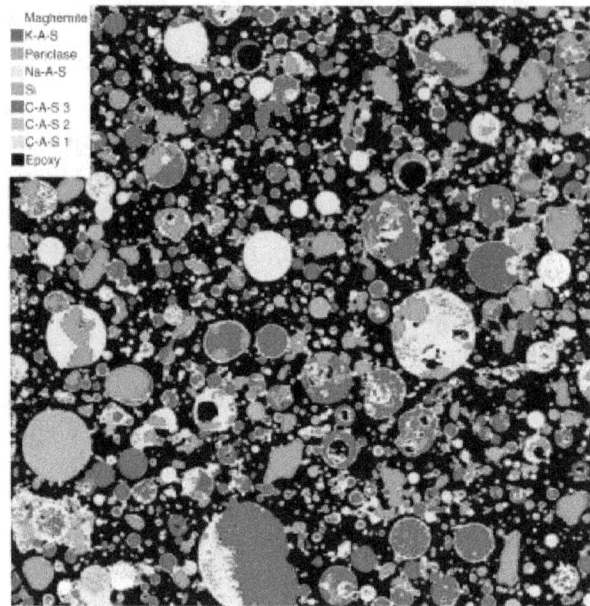

Figure 1. Cluster map showing phase distinction in a Class F fly ash based upon multispectral analysis. Field width = 2050 μm (taken from [14]).

Equally challenging to a proper quantitative characterization of the fly ash is an evaluation of its reactivity in a particular cement-based system. While overall reactivity in an HVFA system can be readily evaluated using calorimetry (as described later) or other established techniques, determining the individual degrees of reaction of the cement and fly ash (as well as phases within the fly ash) remains a challenging task. Two approaches that have been applied by various research groups in the laboratory include microscopy-based analysis [16,17] and selective dissolutions [18-21], the latter typically employing a water-methanol-picric acid solution. The microscopy-based techniques are based on the principle that the unreacted fly ash particles can be readily identified in the backscattered electron image (BEI) of a polished specimen of hydrated blended cement paste. Once the amount of remaining unreacted fly ash is quantified using point-counting [16] or automatic image processing methods [17], its degree of reaction can be computed based on knowing the initial volumetric proportions of the mixture. In selective dissolution [17-21], the reaction products and unhydrated (partially hydrated) cement particles are dissolved and the remaining unreacted fly ash is thus quantified. These techniques have been developed and applied in the laboratory and have not yet been reduced to common practice for field specimens, etc.

Relevance to Industry Practice:
1) While the deficiencies of the current ASTM C618 classification scheme for fly ash are generally well-recognized, advanced characterization techniques are still in the development stage and it is unlikely that new standardized test methods will be approved in the near future.

2) Advanced characterization techniques including X-ray diffraction and SEM with X-ray microanalysis, both supplemented by selective dissolutions, can provide a quantitative characterization of the glassy and crystalline phases in a particular fly ash.

3) Techniques have been developed for quantifying the degree of reaction of fly ash in cement-based materials, but to date, these have only been employed in research studies. Further efforts are needed to mainstream these into common practice.

Mixture Proportioning

The American Concrete Institute (ACI) Committee 211 has developed documents that provide specific guidance on proportioning concrete mixtures. For structural concretes, the ACI 211.1-91 guide [22] recommends replacement ranges of 15 % to 25 % and 15 % to 35 % for Class F and Class C fly ashes, respectively. Both of these ranges fall short of what is generally accepted as the fly ash content of an HVFA mixture. Still, the general procedures provided in ACI 211.1 can be used to design trial batches of HVFA concrete mixtures. The ACI 211.1 guide document [22] does recognize the inherent differences in specific gravities between cement and many supplementary cementitious materials (SCMs), and that the yield of these mixtures will need to be adjusted based on the actual specific gravities of the materials used. Normally, this adjustment is based on reducing the sand content in the blended cement concrete mixture. The document provides other cautionary information concerning the utilization of blended cements with respect to air entrainment, setting times, and application during cold weather conditions.

ACI Committee 232 addresses issues specific to the use of fly ash in concrete. Their report ACI 232.2-03 [23] does contain a chapter (less than two pages) on mixture proportioning, but little quantitative information is provided and no special consideration is given to HVFA concrete mixtures. However, the committee is currently working on a report specifically on the topic of HVFA concretes that may provide more specific details and new insights into the proportioning of these sustainable mixtures.

Approaches to mixture proportioning continue to be developed and promulgated, such as the method of concrete equivalent mortar that utilizes mortar mixtures to proportion a concrete with a reduction in the number of required concrete batches [24]. Another more recent example would be the International Center for Aggregate Research (ICAR) 'aggregate suspension' proportioning method for self-consolidating and other concretes, developed by Fowler and Koehler [25]. This method is currently being developed into an ACI 211 technical note, with both mass-based and volumetric-based proportioning procedures being detailed.

Many ready-mix concrete suppliers, particularly the larger ones, have their own in-house system (software, spreadsheet, and/or procedures) for proportioning concrete mixtures to meet specific requirements. These systems are typically developed based on in-house experience with specific materials' suppliers. Therefore, they are not always immediately applicable to other concrete suppliers, and the techniques may be proprietary.

One immediate need is a source of publicly available, generalized approaches to designing HVFA concrete mixtures. As a first step in this direction, some further considerations for proportioning HVFA concrete mixtures to meet specific targeted performance requirements are provided in Appendices A through D of this report.

Relevance to Industry Practice:
1) Numerous mixture proportioning methods have been developed and are commonly used by the concrete construction industry, but few currently provide specific considerations for creating HVFA concrete mixtures.

2) Due to the large difference in specific gravity between cement and most fly ashes, volumetric-based replacements and proportioning may be worthy of consideration in lieu of conventional mass-based approaches.

3) Regardless of the mixture proportioning procedures employed, trial batches should be prepared initially to verify performance, and subsequently, whenever there is a change in the source of any of the raw materials (cement, fly ash, or admixtures). Including isothermal or semi-adiabatic calorimetry measurements as part of the trial batching process can contribute to a faster isolation and resolution of issues, such as materials' incompatibilities (see next section for further details).

Analysis for Incompatibilities

One of the first steps in properly designing an HVFA concrete mixture (see flowchart in Appendix A [26]) is to ensure that the cement, fly ash, and other powders and chemical admixtures being combined together are compatible with one another. This is critically important because an improper combination of materials can lead to mixtures that flash set or conversely sit for hours without setting, the latter case being due to excessive retardation of the hydration reactions. Isothermal calorimetry is a powerful tool for examining the compatibility of these materials, as outlined in the ASTM C1679 Standard Practice for Measuring Hydration Kinetics of Hydraulic Cementitious Mixtures Using Isothermal Calorimetry [27]; a similar standard based on semi-adiabatic calorimetry [28] is currently in the process of being balloted within ASTM.

Sulfate Optimization

Because many fly ashes provide a significant source of aluminate phases and/or their own source of (calcium) sulfates, sulfate optimization is often an issue in producing HVFA concretes that have desired constructability properties [29]. Portland cements are generally optimized with respect to sulfate content based on measured compressive strengths at a specific target age. This "sulfate optimization" is performed under the assumption that the material will be subsequently employed in a 100 % ordinary portland cement (OPC) concrete. The incorporation of relatively large quantities of fly ash in HVFA concretes can thus result in systems that are either under-sulfated, over-sulfated, or in fortuitous cases, still properly sulfated. As an example, Figure 2 and Figure 3 show the influence of additional gypsum on the hydration reactions in two different blended cement/fly ash mixtures prepared using the same (highly reactive) Class C fly ash [30,31]; in Figure 2, a Type II/V cement (2.5 % SO_3 content by mass) was used, while in Figure 3, a Type I/II cement (3.27 % SO_3 content by mass) was employed. As can be seen in Figure 2, for the first cement blended with the fly ash on a 50:50 **mass** basis, the addition of supplemental gypsum has a pronounced positive effect on the hydration reactions, restoring the shape of the isothermal calorimetry curve to its expected form, but not eliminating the excessive retardation produced by this Class C fly ash. While both the 1 % and 2 % addition levels of gypsum produce reasonable calorimetry curves with distinct silicate and aluminate hydration peaks, 2 % was selected as the optimum value for subsequent mixtures, based on its provision of a slightly higher cumulative heat release value at 24 h [30], which should correspond to a higher 1 d strength as well [29]. Conversely, as shown in Figure 3, when the same fly ash is used with a Type I/II cement at a 40:60 fly ash:cement **volumetric** ratio, a 2 % addition of gypsum only results in a further unwanted retardation of the hydration reactions, without significantly altering the shapes of the hydration peaks. In both cases, the influences of additional gypsum on the hydration reactions can be observed by using small samples and standardized isothermal calorimetry experiments on paste specimens, as opposed to the more labor-intensive conventional employment of strength testing [29].

High Range Water Reducing Agent Incompatibilities

While the previous example focused on incompatibilities due to sulfate imbalance in simple cement/fly ash mixtures without any chemical admixtures, additional issues often arise when high range water reducing agents (HRWRAs) and other admixtures are included in the

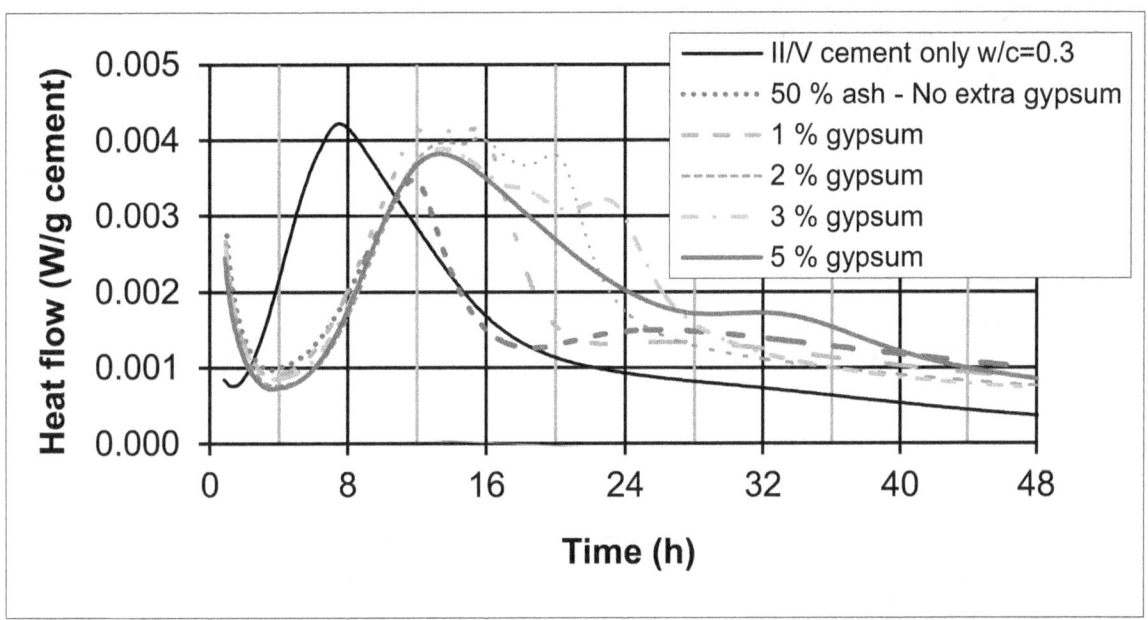

Figure 2. Isothermal calorimetry curves for Type II/V cement/Class C fly ash blends (*w/cm*=0.3) with various addition levels of gypsum [30].

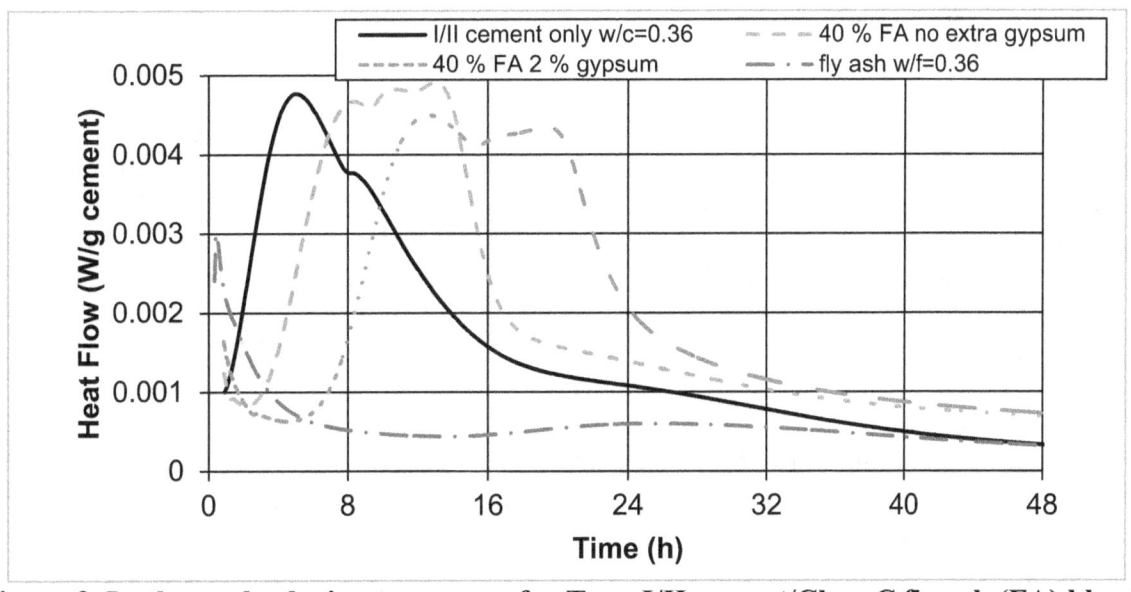

Figure 3. Isothermal calorimetry curves for Type I/II cement/Class C fly ash (FA) blends (*w*/cm=0.36) with and without additional gypsum.

concrete mixtures. As shown by Roberts and Taylor [32], excessive retardation can also be produced by the improper selection of an HRWRA; in their particular study, a lignin-carbohydrate water-reducing admixture was being used. A further example of this influence of HRWRA on performance is provided in Figure 4, showing the difference in hydration response depending on the type of HRWRA employed in a Type II/V cement/Class F fly ash blended paste (50:50 on a mass basis). While both HRWRAs increase retardation, the performance of HRWRA-B would be unacceptable from a practical concrete construction viewpoint. The Class F fly ash by itself produced no retardation, so that in this case, any delays in setting time for a blended paste without HRWRA would be due simply to a dilution effect (the blended mixture containing only 50 % of the cement present in the control mixture), as will be discussed further in a subsequent section.

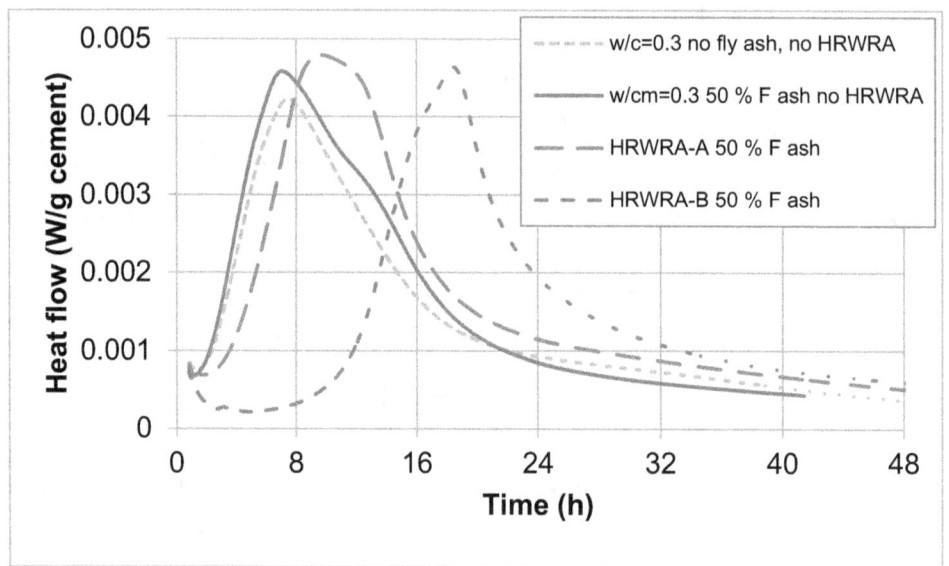

Figure 4. Isothermal calorimetry curves for a Type II/V cement/Class F fly ash blend with (two different) and without an HRWRA.

Calorimetry is a useful tool in developing HVFA mixtures because it provides diagnostic information about the hydration reactions occurring in the paste. As complete and useful as this measurement technique is, it cannot readily predict all the important early-age performance of an HVFA mixture: false setting, early stiffening, etc. A more complete picture of how these mixtures will perform will require additional test methods, such as the modified versions of the ASTM C359-08 early stiffening test, a mini-slump test (currently under ASTM ballot), and rheometer tests based on the guidance provided in ASTM C1749-12 [32,33].

Relevance to Industry Practice:
1) Isothermal and/or semi-adiabatic calorimetry (the latter as demonstrated in reference [28]) provide valuable rapid screening methods for identifying potential incompatibility issues in HVFA concrete mixtures.

2) A cement whose sulfate content has been optimized by its manufacturer may not contain optimal sulfate when blended with fly ash (and other powders) in an HVFA concrete mixture.

3) Chemical admixtures, particularly HRWRAs, can also play a critical role in the early-age performance (including incompatibilities and retardation issues) of HVFA concrete mixtures.

Curing Options

Because most fly ashes react at a much slower rate than portland cement, concrete mixtures containing significant amounts of fly ash will often benefit from extended curing beyond that conventionally employed for 100 % OPC concretes. These HVFA concretes are, therefore, also more sensitive to the quality of curing conditions [34,35], as mix water that is not consumed in early-age chemical reactions or absorbed by early-age hydration products may evaporate if the concrete is exposed to a drying environment. Applying measures to prevent this evaporation, via curing membranes or curing compounds for example, must be extended beyond the timeframe typically used with OPC concrete to provide additional time for the cement and fly ash reactions to occur, which will further densify and strengthen the concrete mixture. If misting, fogging, or wet burlap are employed, in addition to preventing evaporation from the concrete, an additional source of external water becomes available to counteract the self-desiccation that accompanies the chemical reactions occurring within the paste portion of the HVFA concrete.

However, not all concrete mixtures will benefit equally from additional external water. In lower *w/cm* concretes (e.g., at or below 0.4), as the hydration and pozzolanic reactions occur, the capillary pores within the microstructure will disconnect (depercolate) so that the imbibition rate of external water will be significantly inhibited, resulting in self-desiccation that will become significant at depths just a few centimeters below the external water surface. This self-desiccation is accompanied by the generation of internal stresses that may contribute to early-age cracking of the concrete. And, as a practical matter, supplying an external water source for vertical surfaces is much more difficult than doing so for horizontal ones.

In the past decade, to address these issues, a new paradigm for curing concrete has emerged, namely internal curing (IC) [36]. In IC, additional water is supplied to the hydrating paste via the incorporation of water reservoirs into the concrete mixture. Most often, pre-wetted lightweight aggregates (LWA) are used as the reservoirs, but superabsorbent polymers (SAPs) and pre-wetted wood fibers have also been employed in some laboratory studies. This approach works by distributing these small water reservoirs uniformly throughout the three-dimensional structure and curing the concrete from the inside out. The water transport distances are thus minimized and the hydrating paste microstructure remains saturated throughout, even as the capillary porosity depercolates. This leads to a reduction in early-age cracking and an improvement in longer term mechanical and transport properties, the latter being mostly due to enhanced reactions of the cement and fly ash and the high quality interfacial transition zone that is developed between paste and LWA [36].

The efficacy of IC for HVFA mixtures has been examined recently in a study where the properties of mortar mixtures with 40 % and 60 % Class C fly ash (volumetric replacement for cement) with and without IC were contrasted [37]. In that study, IC was supplied via a pre-wetted fine LWA that replaced a portion of the normal weight sand in the examined mortars. The performance target was a *w/c*=0.42 100 % OPC mortar and as is common practice, the *w/cm* of the HVFA mixtures was reduced to 0.3 to obtain concretes with acceptable early-age strengths. Measured compressive strengths vs. time for the various mixtures are reproduced here in Figure 5. For the particular materials employed in this study, the benefits of the additional hydration provided by the IC outweighed the incorporation of (typically weaker than sand) fine LWA into the mixtures, so that compressive strengths gains were observed when comparing similar mixtures (40 % or 60 % fly ash) with and without IC. The contribution of fly ash to longer term strength is also seen in Figure 5, as both of the mixtures with 40 % fly ash exhibited a higher strength than the w/c=0.3 OPC control mixture for ages of 28 d and beyond. Even at the 60 % replacement level, the mixture with IC exhibited a higher strength than the w/c=0.3 control at ages of 91 d and beyond. All HVFA mixtures exhibited a higher strength than the target w/c=0.42 mixture for ages of 14 d and beyond. While IC produced an increase in compressive strength, results for elastic modulus (not shown here) indicated a reduction on the order of 10 % when HVFA mortar mixtures with IC were compared to their non-IC counterparts [37]. A beneficial side effect of a reduced elastic modulus, however, could be a reduced cracking tendency in these HVFA mixtures with IC.

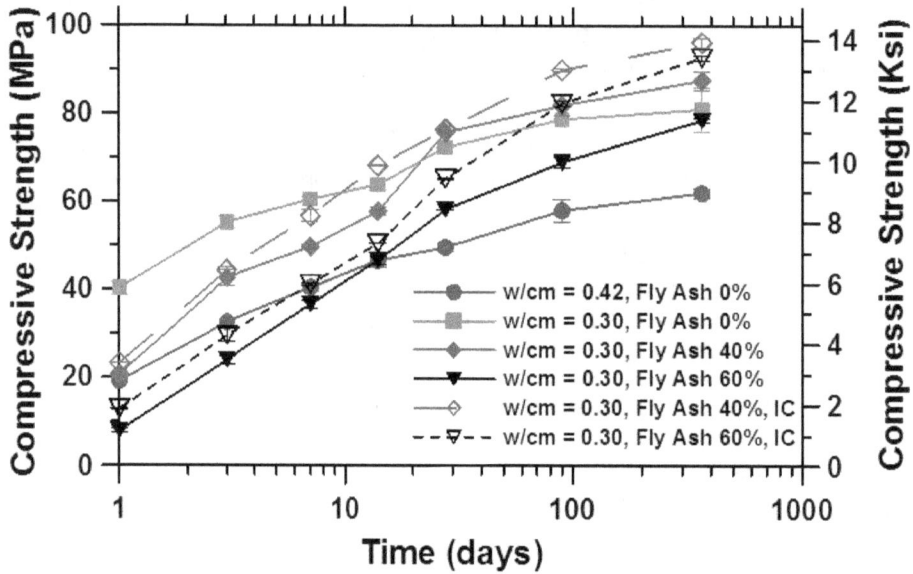

Figure 5. Compressive strengths of HVFA mortar mixtures with and without IC (taken from [37]). Error bars represent the standard deviation from the average of three samples.

Indeed, one of the main reasons to implement IC in practice is to reduce early-age (autogenous) shrinkage and the cracking that often accompanies it. Figure 6 shows the measured autogenous shrinkage of sealed cylindrical specimens during the first 14 d of curing for the various mortars evaluated in the study [37]. While the HVFA mixtures without IC exhibit less shrinkage than the w/c=0.3 OPC mortar at ages out to 14 d, due to the fly ash being less reactive than the cement (dilution effect) during this time period, the HVFA mortars with IC exhibit a deformation that is quite similar to that of the w/c=0.42 target OPC mortar. Corresponding sealed (cracking) ring specimens were prepared for a subset of these mortars to evaluate their stress development and cracking behavior. As shown in Figure 7, surprisingly, the only specimen to exhibit cracking during the 28 d test was the mortar with 40 % fly ash and without IC, which develops significant autogenous shrinkage at intermediate ages (14 d to 28 d). The corresponding mortar with IC exhibited a stress development curve quite similar to that of the target w/c=0.42 OPC mortar, with no evidence of cracking over the course of the 28 d evaluation period.

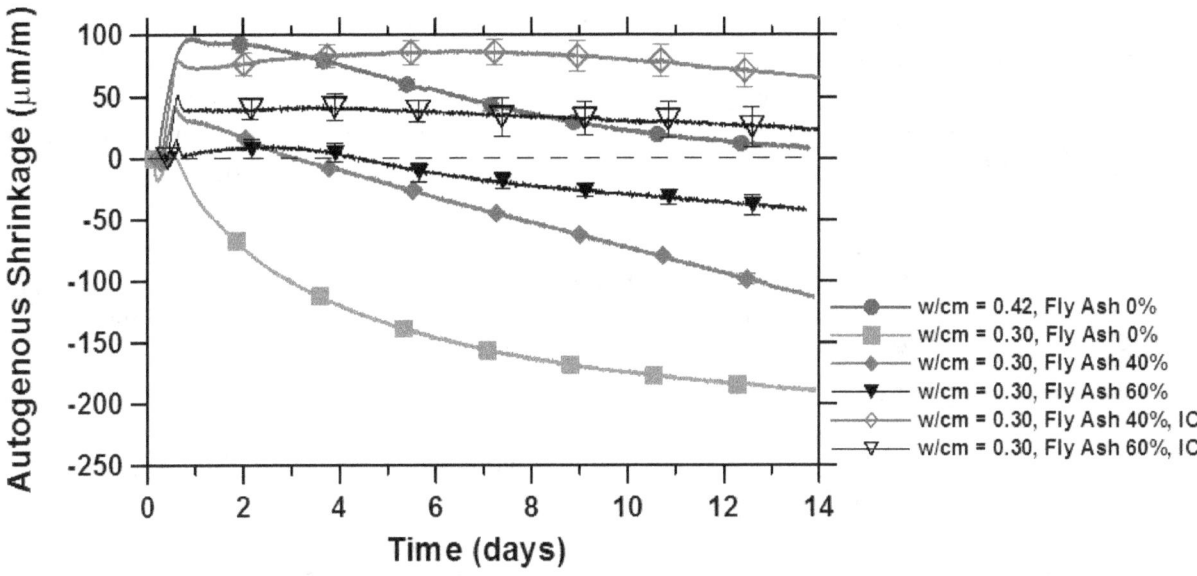

Figure 6. Autogenous shrinkage as a function of time for HVFA mortar mixtures with and without IC (taken from [37]). Error bars represent the standard deviation from the average of three samples.

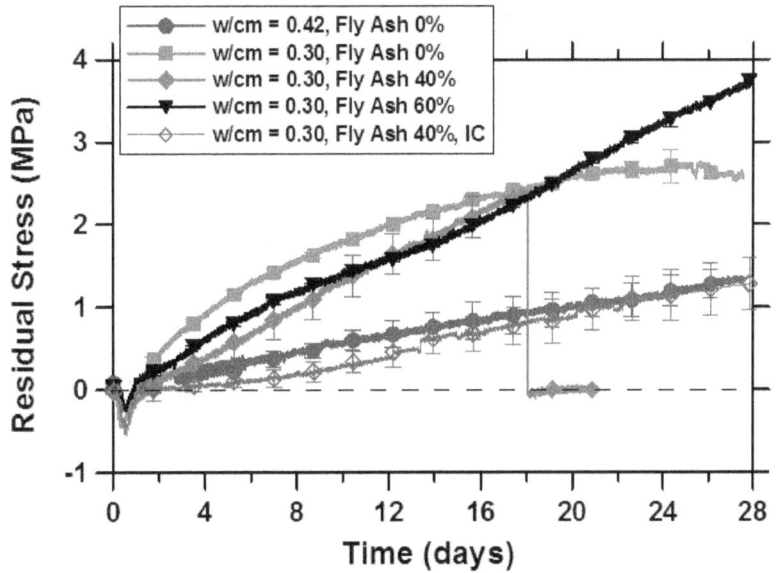

Figure 7. Stress developed in ring specimens as a function of time for sealed mortar mixtures (taken from [37]). Error bars represent the standard deviation from the average of two samples.

While it is recommended that internal curing be accompanied by external measures, such as application of a curing compound or a curing membrane, to minimize evaporation, next generation self-curing concretes (requiring no external curing) have entered the marketplace as of early 2012 in Europe and elsewhere. As outlined by Mather in 2001 [38], many concretes are effectively self-curing, as they receive favorable environmental (weather) conditions following their placement that promote hydration and strength development, without the necessity for external curing measures. In the same article, Mather mentions several practical examples of the ability of LWA concretes to provide additional curing water to the concrete mixture when the aggregates are batched wet, once again contributing to a 'self-curing' concrete. In 2012, a major international concrete ready-mix supplier announced the release of a new global brand of self-curing concrete, with the self-curing properties being achieved through a combination of unique mix design principles and proprietary admixtures. With the overall push towards sustainability, it seems likely that HVFA counterparts of these self-curing mixtures will soon follow.

Relevance to Industry Practice:

1) In general, the reactions between the fly ash and the cement clinker and its hydration product phases proceed more slowly than the cement hydration reactions themselves. Thus, HVFA concretes will typically benefit from an extended period of controlled (moist) curing beyond the 'conventional' curing period of 7 d.

2) When HVFA mixtures use reduced *w/cm* and internal curing, they can achieve strength and stability properties similar to OPC mixtures prepared at more moderate *w/c*.

3) One method for maintaining a higher moisture level (internal relative humidity) within a field HVFA concrete is to employ the technology of internal curing.

4) Self-curing concretes are emerging as commercially available products and new self-curing HVFA concretes will likely emerge in the coming years.

Workability

One of concrete's most important properties is workability, as the concrete has to be placed and consolidated to ensure that the hardened properties will be achieved as designed. A concrete that is not well consolidated or not able to be easily placed, will not have the intended durability or other requisite hardened properties. Workability is defined by ACI as "that property of freshly mixed concrete or mortar that determines the ease with which it can be mixed, placed, consolidated, and finished to a homogenous condition" and is usually measured or specified by slump for conventional concrete or by slump flow for self-consolidating concrete (SCC). These properties, however, depend upon a number of factors, and it can be difficult to correlate changes in mixture proportions to changes in slump and/or slump flow.

At a more fundamental level, the flow properties that would ensure a good concrete workability are the rheological properties of the material. Conventional concrete flow is commonly characterized using a Bingham model that assumes that a finite stress (the yield stress) is needed to initiate flow, and that additional stress results in additional strain (flow) according to the material's plastic viscosity (the change in stress with strain). For SCC, often the Hershel-Bulkley model is used instead, in which the relationship between shear stress and shear rate includes the shear rate raised to a power, n ($n=1$ reduces to the Bingham model). Thus, the rheological properties of concrete are mostly characterized using two parameters: its plastic viscosity and its yield stress as defined by the Bingham equation.

Determining the yield stress and the plastic viscosity, in practice, remains a challenge. Several studies have shown that the yield stress is correlated with the slump value [39,40,41]. In contrast, plastic viscosity can only be determined through the use of a rheometer. There is, for now, no standard concrete rheometer and thus the rheological values need to be reported while specifying the specific concrete rheometer being employed [42,43]. NIST is working in collaboration with ACI Committee 238 to develop a Standard Reference Material (SRM) for the calibration of the concrete rheometers. While for now only the SRM for paste is available (SRM 2492), it is expected that the mortar SRM will be ready for release in 2013 and the concrete SRM in 2014.

The workability or rheological properties are influenced by the composition of the mixture, including cement type and particle size distribution (PSD), SCMs' type and PSD, aggregates' shape and size distribution, water/cement ratio and addition of chemical admixtures (HRWRA), and air content. Addressing the impact of all of these factors is beyond the scope of this guide. Instead, this section will only summarize some select results and provide some guidelines on optimizing the flow when fly ash is used as a replacement for cement.

As mentioned previously, the addition of fly ash is in practice commonly performed by mass replacement of cement. This practice results in a significant change in the proportioning of the whole concrete, because the density of fly ash (typically in the range of 2100 kg/m^3 to 2700 kg/m^3) is significantly lower than that of cement (about 3150 kg/m^3). When performed as a mass replacement, the paste volume will change as a result of scaling the mixture to achieve the desired yield. Therefore, changes in rheology will depend on both the variable paste volume fraction and the properties of the fly ash. Thus, it would be preferable to determine the optimum replacement of the cement by volume, thus keeping the paste volume fraction at a constant value. In the literature, however, most if not all of the investigations are performed utilizing a mass replacement.

The addition of fly ash improves workability, but also increases stability or reduces segregation, while reducing the amount of HRWRA needed to achieve a desired flow [44,45]. This quality is attributed to the spherical shape of the fly ash particles that would operate as ball bearings [46], or improving the packing density of the cementitious materials [47]. Nevertheless, not all fly ashes improve fluidity to the same degree. Ferraris et al. [44] showed that a mean particle size of 3 μm could improve fluidity, while if the mean particle size was 5.7 μm, the fluidity could degrade, as compared to that of the plain cement paste. Similar conclusions were reached by Lee et al. [47] where the fluidity of a cement paste with fly ash increases with the width of the particle size distribution of fly ash as characterized by the factor n of a Rosin-Rammler distribution function. Kwan and Li [48], employing a fly ash microsphere (FAM) addition that is a superfine fly ash, found that fluidity is improved both in cement paste and in mortar. They attribute the increased fluidity to the increased packing density, as the FAMs are smaller than cement. As the paste fluidity is increased, the amount of paste needed in mortar to achieve the same flow is reduced. This introduced a concept that the addition of fly ash affects mainly the cement paste portion, thus it is advocated that to determine the incompatibility of the fly ash with HRWRA and the cement, or to optimize the amount of fly ash, cement paste measurements would give an optimal composition without the need for lengthy trials with concrete and mortar [44,48,49]. In addition to the ball-bearing and particle packing effects, results have also indicated that fly ash may contribute to workability improvements by decreasing the flocculation of the cement particles (via a particle dilution effect) [50].

Self-consolidating concrete (SCC) is a type of concrete that flows by gravity and does not need consolidation by vibration. To achieve the desired flow properties, the material needs to be stable (no segregation during flow even in structures with a high density of rebar) and possess a zero yield stress. To achieve these properties there are various solutions: increase water content, increase HRWRA, increase fines content by increasing paste content (e.g., limestone fines addition), and including viscosifiers. Obviously, simply increasing the water content could lead to reduced durability and strength, and is thus not the first choice. Increases in HRWRA and/or viscosifier can be expensive possible solutions. Another more economical possibility is to increase the fly ash replacement that would increase flow while decreasing the HRWRA dosage needed [44]. This is often the desired solution as it has also other benefits: more cost effective, increased durability, and reduced heat of hydration, when compared with an increase in cement content alone.

Specifications usually limit the amount of cement replacement, but some studies have reported beneficial properties, especially with respect to workability, with up to 30 % replacement by volume [45,46], and with up to 50 % replacement exhibiting a concurrent reduction of bleeding and aggregate blockage [51].

Relevance to Industry Practice:
1) In general, fly ash replacement of cement improves workability and reduces the need for HRWRA, even at high replacement rates of fly ash for cement.

2) Optimization of the fly ash amount, type and PSD can be done by testing cement paste instead of concrete or mortar, with significant savings in trial batches and material and capital resources.

3) The usage of fly ash increases stability, reducing bleeding and segregation in concretes such as SCC.

Setting Time

As cited in the 2012 NRMCA survey [2], a common issue encountered with HVFA concrete mixtures is their delayed setting, with corresponding delays in finishing and other construction operations.[2] Due to a substantial research effort in recent years, several viable mixture proportioning options have been developed to mitigate excessive delays in setting for HVFA concrete mixtures.

As with conventional OPC concretes, non-chloride accelerators can be effectively employed to reduce setting times of HVFA concrete mixtures [52,53]. Higher dosages may be required to attain target setting times (such as that of a comparable 100 % OPC concrete mixture) depending on the characteristics of the fly ash, the replacement level, and the expected curing temperatures. In general, lower curing temperatures and higher CaO contents in Class C fly ashes require higher dosages and, in some cases, acceptable setting behavior of a 50 % HVFA mixture may not be achievable via this approach [52,53]. Non-chloride accelerators also can be fairly expensive, so that economic considerations may sometimes dictate the viability of their utilization in a proposed HVFA mixture.

As will be shown in the section to follow, one convenient method (that is already employed in practice, particularly during the winter construction season) for increasing the early-age strength of HVFA concrete mixtures is to switch to an ASTM C150 Type III (finer) portland cement. Such a change typically has a significantly greater effect on early-age strength development than on initial and final setting times. For example, one recent study [26] indicated that switching to a Type III cement reduced the setting time delays by only about 1 h for two HVFA mortars (one with a Class C fly ash and the other with a Class F) that originally exhibited a 3 h to 4 h delay in setting relative to a corresponding 100 % OPC control mortar. For these same two HVFA mortars, this switch to Type III cement increased their 1 d compressive strengths by about 60 % on average. If Type III cement is being employed to enhance these early-age strengths, its potential contributions to reducing setting time delays should also be taken into consideration as the mixture proportions are being developed and optimized.

Various powder additions have been investigated for their ability to restore setting times of HVFA mixtures to values typical of those achieved for corresponding 100 % OPC control mixtures. One study screened numerous candidate powders and identified calcium hydroxide and a rapid set cement as the two most promising candidates [30]. Representative setting times resulting from these additions in HVFA pastes prepared with either a Class C or a Class F fly ash are provided in Table 2 [54]. For both fly ashes, a 5 % calcium hydroxide addition by mass restored the initial and final setting times to be nearly identical to those of the control (no fly ash) cement paste. For the rapid set cement, while a 5 % addition was adequate for the Class F fly ash mixture, a 10 % addition was required for the Class C fly ash. One additional consideration in employing these powder additions is their subsequent influence on compressive strength. While the rapid set cement addition has produced mortars with similar or superior long term (28 d or 91 d) strengths to those of the control 100 % OPC mortar, a significant decrease in 28 d strength on the order of 15 % has been observed for HVFA mortars prepared with Class F fly ash and a 5 %

---

[2] Low early-age strengths that also influence the timing of construction operations, such as formwork removal, will be considered as a separate issue in a subsequent section of this report.

calcium hydroxide addition [26,30]. Still, calcium hydroxide additions have been recently employed in practice to produce HVFA concretes (70 % replacement of OPC) with greater bonding to reinforcing steel than those of the 100 % OPC control concrete mixtures [55], although the measured compressive strengths of the HVFA mixtures were indeed about 20 % lower than that of the control concrete.

Table 2. Setting Times for HVFA Paste Mixtures [54]

Paste mixture	Vicat initial set (h)	Vicat final set (h)
Type II/V cement 0.67 % HRWRA	5.1 h	5.9 h
50 % C ash, 2 % gypsum, 0.33 % HRWRA	8.2 h	8.8 h
50 % C ash, 2 % gypsum, 5 % CH, 0.33 % HRWRA	5.3 h	6.0 h
50 % C ash, 2 % gypsum, 10 % rapid set cement, 0.33 % HRWRA	3.1 h	4.5 h
50 % F ash, 0.87 % HRWRA	8.6 h	10.2 h
50 % F ash, 5 % CH, 0.87 % HRWRA	5.2 h	5.9 h
50 % F ash, 5 % rapid set cement, 0.87 % HRWRA	3.3 h	4.5 h

In the initial screening study described above, a relatively coarse limestone powder, with a modal particle diameter of about 27 μm, had basically no effect on the early-age hydration response as assessed using isothermal calorimetry [30]. However, subsequent investigations have indicated that limestone additions can be quite effective for mitigating setting time delays (and increasing early-age strengths) in HVFA mixtures, when finer limestone particles are employed [31,56-61]. Results when employing a fine limestone with a median particle diameter on the order of 1 μm have been particularly promising [59,61]. The fine limestone particles provide additional surface area for the precipitation and growth of calcium silicate hydrate gel and other products from the cement hydration and pozzolanic reactions. Additionally, the fine limestone particles can also participate in these reactions leading to the formation of stable carboaluminate (as opposed to conventional sulfoaluminate) phases. The carboaluminate phases may be stiffer than the sulfoaluminate phases that would be formed in the absence of limestone, further contributing to strength enhancements [62].

A study employing four different modal diameters of limestone powders has indicated that the reduction in setting time for a fixed set of mixture proportions is directly proportional to the provided limestone surface area (as quantified by BET surface area measurements) [59]. Figure 8 illustrates this linear relation for mixtures with either a Class C or a Class F fly ash, where in some mixtures a portion of the fly ash has been replaced by limestone powder. In this study, titanium dioxide (anatase) was also included, as an inert material with a surface area similar to that of one of the limestone powders, to confirm that both the size and chemical nature

21

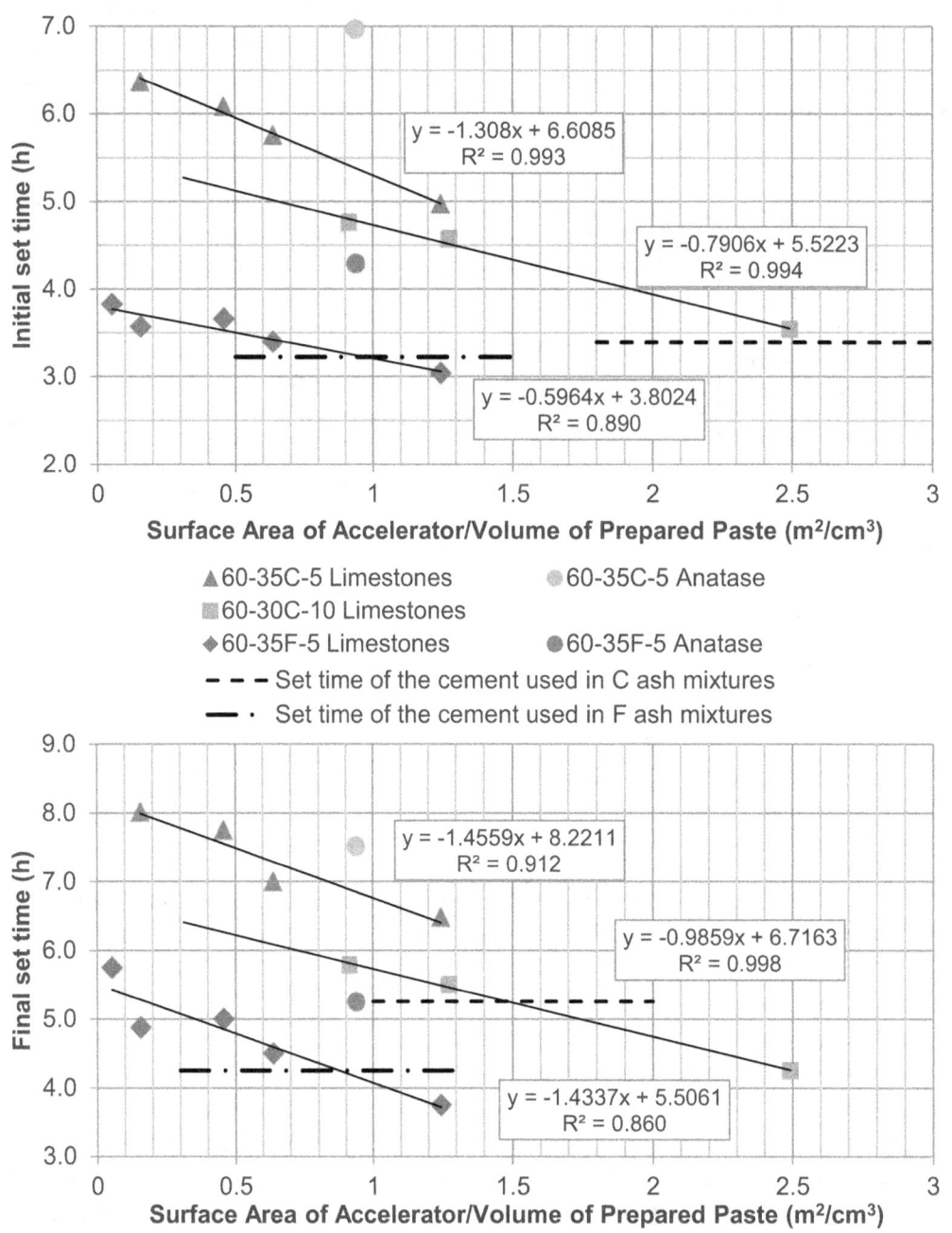

Figure 8. Correlation between initial (top) and final (bottom) setting times and surface areas of the accelerating powders (taken from [44]).

of the powder material are critical to its ability to accelerate early-age reactions in cement-based materials. While a nanolimestone provided the most significant reduction in setting times, the performance of a fine limestone powder with a 0.7 μm median particle diameter was also quite acceptable. This latter fine limestone has been subsequently employed in a series of HVFA concrete mixtures [61], whose mixture proportions are provided in Table 3. As indicated in Figure 9, the replacement of a portion of the fly ash with a fine limestone powder continues to markedly reduce setting time delays in these more sustainable concretes. As will be discussed subsequently, these fine limestone replacements for fly ash also produce significant improvements in compressive strength and transport (electrical) properties.

Table 3. Concrete Mixture Proportions for Study of HVFA with Fine Limestone [61]

Mix ID	Cementitious (kg/m³) [(lb/yd³)]	Type I/II cement (kg/m³) [(lb/yd³)]	Class F fly ash (kg/m³) [(lb/yd³)]	Class C fly ash (kg/m³) [(lb/yd³)]	Limestone 0.7 µm (kg/m³) [(lb/yd³)]	Coarse aggreg. (kg/m³) [(lb/yd³)]	Fine aggreg. (kg/m³) [(lb/yd³)]	Water content (kg/m³) [(lb/yd³)]	w/cm	HRWR (fl oz/ cwt)
PC	335 [564]	335 [564]				1040 [1750]	858 [1444]	134 [226]	0.40	7.7
40F	291 [491]	201 [338]	91 [153]			1040 [1750]	858 [1444]	134 [226]	0.46	3.8
30F10L	297 [499]	201 [338]	68 [114]		28 [47]	1040 [1750]	858 [1444]	134 [226]	0.45	3.8
40C	310 [522]	201 [338]		109 [183]		1040 [1750]	858 [1444]	134 [226]	0.43	3.0
30C10L	311 [523]	201 [338]		82 [138]	28 [47]	1040 [1750]	858 [1444]	134 [226]	0.43	3.0
60F	270 [454]	134 [226]	136 [229]			1040 [1750]	858 [1444]	134 [226]	0.50	3.8
45F15L	278 [467]	134 [226]	102 [172]		41.6 [70]	1040 [1750]	858 [1444]	134 [226]	0.48	3.8
60C	298 [501]	134 [226]		163 [275]		1040 [1750]	858 [1444]	134 [226]	0.45	3.0
45C15L	298 [502]	134 [226]		122 [206]	41.6 [70]	1040 [1750]	858 [1444]	134 [226]	0.45	3.0

Relevance to Industry Practice:

1) Similar to their role in OPC concretes, non-chloride accelerators (chemical admixtures) may be employed to reduce the setting time delays commonly encountered in HVFA concrete mixtures. Higher dosages than those typically employed in OPC mixtures may be necessary and costs may become prohibitive.

2) Switching to a Type III cement will typically provide a small improvement (reduction on the order of 1 h) in setting times. The Type III cement will have a much larger positive influence on early-age compressive strengths than on setting time reductions.

3) Calcium hydroxide and rapid set cement powder additions (at a 5 % to 10 % level by mass) can be employed to mitigate retardation in setting times, although the former has a generally negative impact on 28 d compressive strengths.

4) Fine limestone with a median particle diameter on the order of 1 µm as a replacement for a portion of the fly ash in an HVFA mixture is highly efficient in accelerating the cement hydration and pozzolanic reactions, and thus in restoring the setting times of these HVFA concretes to be in the vicinity of those of comparable OPC concrete mixtures. Further benefits (to be presented in subsequent sections) are seen in strength increases and improvements in transport and electrical properties such as RCPT and surface resistivity.

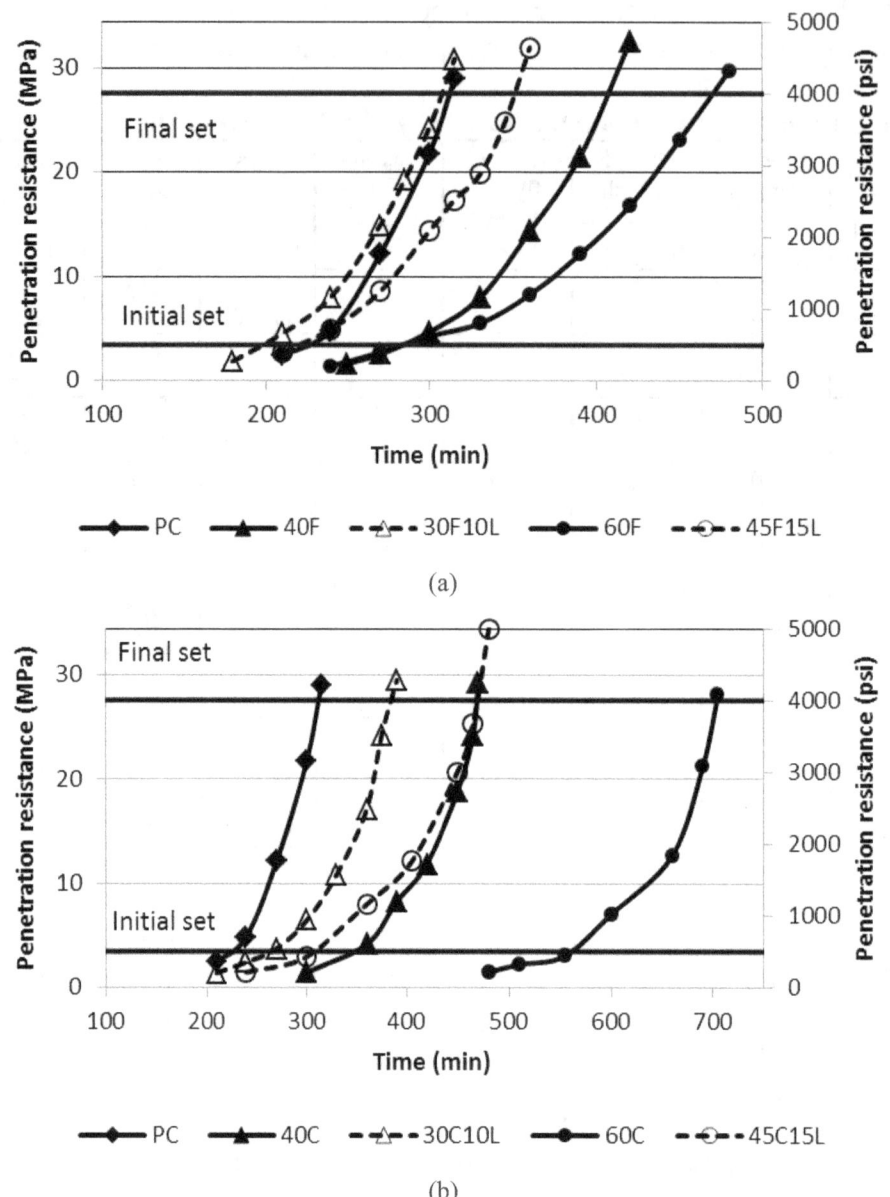

(a)

(b)

Figure 9. Setting development: a) OPC mixture and mixtures containing Class F fly ash
(45F15L indicates a mixture with 45 % Class F fly ash and 15 % limestone by volume); b)
OPC mixture and mixtures containing Class C fly ash (30C10L indicates a mixture with
30 % Class C fly ash and 10 % limestone by volume) (taken from [61]).

24

Strength Development

Based on the seminal works of Abrams [63] and Feret [64], the classic approach to increasing concrete strengths at all ages is to reduce the w/c in an OPC concrete, or equivalently the w/cm in an HVFA concrete. The advent of high range water-reducing agents, also known as superplasticizers in their early days, has permitted substantial reductions in w/c or w/cm, without a significant loss of slump or workability. Thus, decreasing w/cm of an HVFA mixture to obtain more acceptable early-age (and even 28 d) strengths is common practice. In practice, for starting w/c in the range of 0.4 to 0.45 for a 100 % OPC target concrete, typical reductions in transitioning to an HVFA mixture are on the order of 0.05 to 0.10 [26,37,53,61]. However, as noted in Table 1 [6], when the starting w/cm of the target concrete is on the order of 0.3 (e.g., for a high-performance concrete), the w/cm of an HVFA mixture that performs similarly may require less or no reduction. For some cement-fly ash combinations, the reduction in w/cm necessary to meet targeted 28 d strength levels will still be insufficient to provide adequate early-age strength for timely formwork removal and other construction operations. In these cases, further measures to specifically increase early-age strengths may be necessary.

One such measure that has been employed in practice in HVFA concretes is the use of an ASTM C150 Type III cement, which is more finely ground and may have an increased sulfate content by comparison with an ASTM C150 Type I/II/V cement. Representative compressive strength results from a study on HVFA mortars are provided in Figure 10 [26]. Using the strengths of the w/c=0.4 100 % OPC mortar as the target values, the original HVFA (50:50 by mass) mixtures prepared with a reduced w/cm=0.3 fail to achieve the target strengths at 1 d, with only the Class C fly ash mixture reaching the target strength at 7 d. However, switching to a Type III cement brought both fly ash mixtures to an acceptable strength level at 1 d and provided strengths that exceeded the target values at ages of 28 d and beyond. Figure 10 is also informative in that the 365 d strength values of the HVFA mixtures are approaching the levels of a 100 % OPC w/c=0.3 mortar (that itself exhibits a minimal strength increase beyond 28 d), illustrating the propensity of the longer term pozzolanic reactions to make a significant contribution to later age strength development in these HVFA systems, and supporting the transition to later ages (56 d or 91 d) for compliance strength testing of HVFA mixtures, as opposed to the standard age of 28 d commonly employed for OPC concretes.

In addition to their mitigation of excessive setting time delays, replacement of a portion of the fly ash in an HVFA concrete mixture with a fine limestone powder will also have a beneficial influence on compressive strength values. Figure 11 shows measured compressive strengths for HVFA concretes with and without fine (1 µm) limestone replacing a portion of the fly ash on a volume for volume basis [61], with all mixtures prepared with constant volume fractions of water, powders (including cement, fly ash, and limestone), and fine and coarse aggregates. As shown previously in Figure 9, these limestone replacements reduced the setting times of the HVFA mixtures to be nearly equal to those of the control OPC mixture. With respect to strength, at early ages, for both the Class F and the Class C fly ash studied, the fine limestone increases strengths by about 2 MPa (about 300 psi). By 28 d, these strength enhancements have increased significantly, and in the best case, a Class C fly ash HVFA concrete with an initial 28 d strength of 13 MPa (1900 psi) has nearly doubled to 25 MPa (3600 psi). This strength improvement is likely due to a combination of the acceleration provided by the fine limestone powder (surfaces) and the enhanced formation of (stiffer) carboaluminate phases [62] in this particular cement/Class C fly ash/fine limestone ternary blend. While significant, these increases still failed to achieve the target strengths of the OPC concrete. Thus, other measures such as a *w/cm* reduction and/or substitution of a Type III cement would need to be employed along with the fine limestone replacements for fly ash [65].

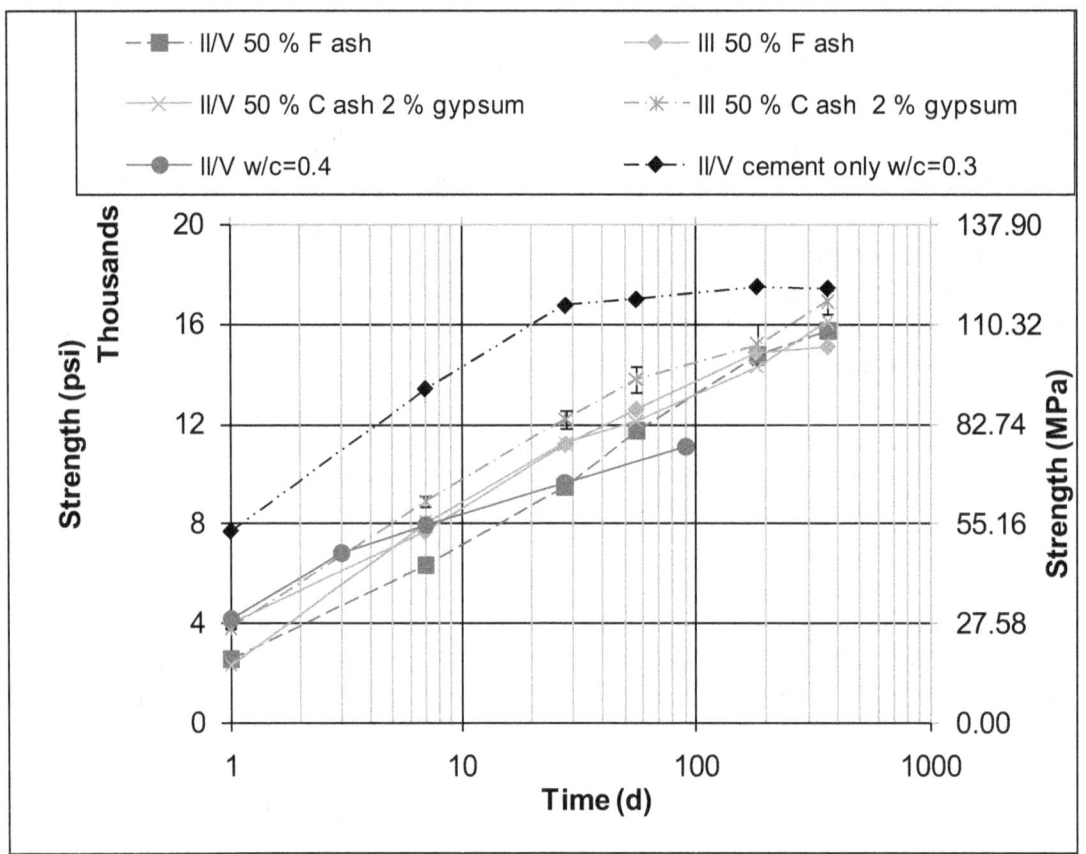

Figure 10. Measured mortar cube compressive strengths vs. age. For the mixtures with fly ash, *w/cm*=0.3 in every case. Error bars (one standard deviation among three specimens) are provided for the III 50 % C ash 2 % gypsum data to provide an indication of variability (taken from [26]).

In a study performed for the U.S. Department of Energy, Obla et al. [66] have examined the applicability of the well-known maturity method to HVFA concrete mixtures. They determined that the maturity method, as conventionally employed to predict strength development in OPC concretes [67,68], is equally applicable to HVFA concretes. Some additional key points of their study included that 1) temperature-matched cured (field) cylinders will generally achieve higher compressive strengths than standard and field-cured cylinders due to the higher curing temperatures produced in (mass) concrete elements, 2) pull-out testing exhibited excellent correlation with cylinder compressive strengths and could thus be employed for in-place field evaluations of strength, and 3) HVFA mortar cubes exhibited higher long-term strengths when cured at higher temperatures versus those cured at standard temperatures. This last observation is contrary to what has been conventionally observed for OPC systems [67] and is thus worthy of further research in the future.

26

When adjusting mixture proportions to alter early-age compressive strengths, the general relation between heat release and strength development, for various curing ages, can be used to good advantage. The example data sets shown in Figure 12 and Figure 13 illustrate this correlation for a wide variety of mortars and concretes, respectively [69]. The mortars were all prepared using silica sand, while the concretes were prepared using a limestone coarse aggregate and silica sand. The fitted lines in these plots can be used to adjust the *w/cm* of an HVFA mixture to obtain a desired compressive strength level, based on the measured compressive strength of an existing trial mixture. Further details on applying this approach to mixture proportioning modification are provided in Appendices B through D of this guide.

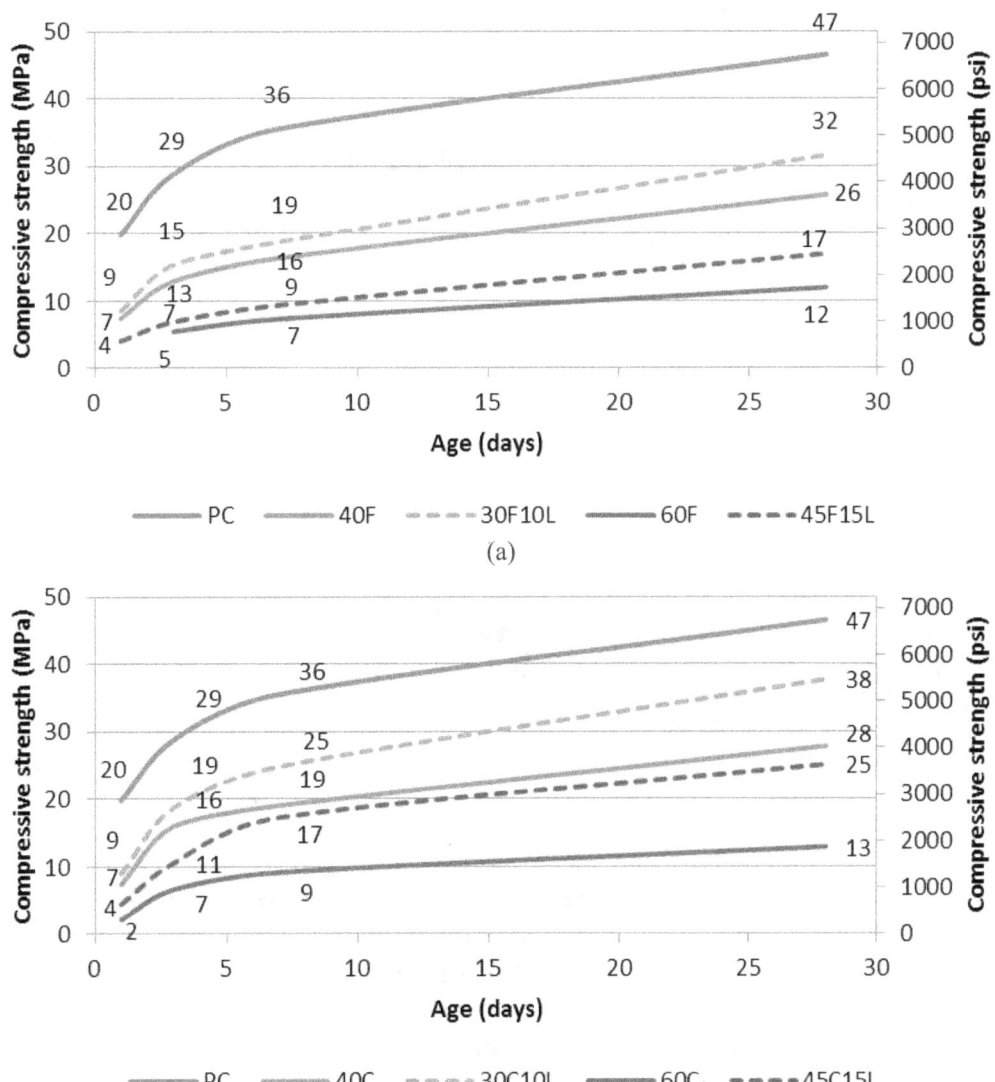

Figure 11. Compressive strength development for concrete mixtures (see Table 3) containing: a) Class F fly ash and b) Class C fly ash (taken from [61]). Coefficients of variation for three replicate specimens varied from 0.67 % to 2.5 % for the various mixtures. Numbers above lines indicate strengths obtained at ages of 1 d, 3 d, 7 d, and 28 d.

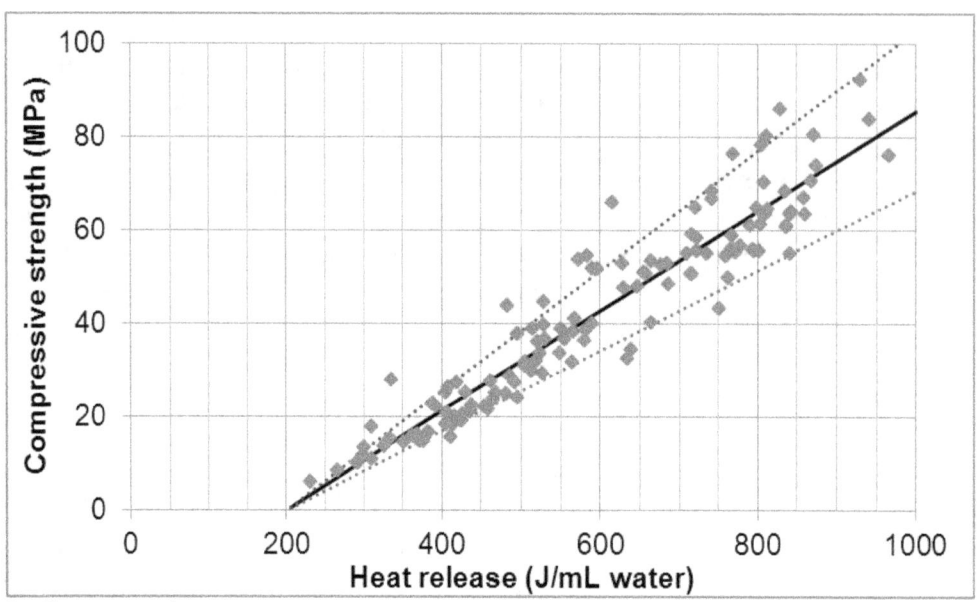

Figure 12. Compressive strength vs. heat release per mL of water for mortars evaluated at a variety of ages (taken from [69]).

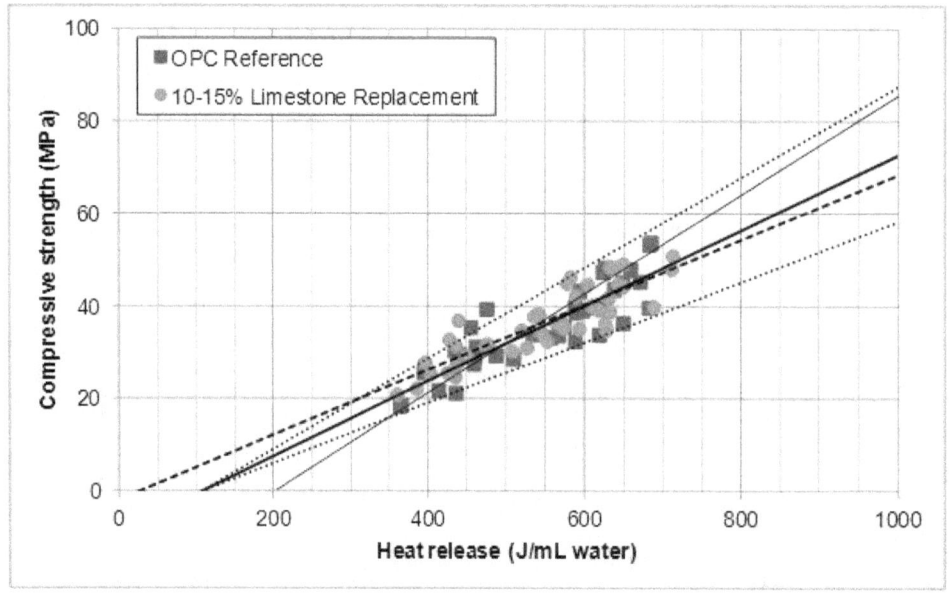

Figure 13. Compressive strength of concrete cylinders vs. heat release per mL of water for mortars from [70]. The bold solid line indicates the best fit linear relationship (R^2=0.77) for the OPC concretes, with the two dotted lines indicating ± 20 % from the best fit values. The dashed line represents the best fit for the limestone replacements (R^2=0.76). The thin solid line represents the best fit determined for the mortar data in Figure 12. (Taken from [69]).

Relevance to Industry Practice:

1) To achieve acceptable early-age strength, HVFA concretes are commonly proportioned with a lower *w/cm* than that of the target (100 % OPC) concrete. Because of the spherical shape of many of the fly ash particles and the considerable dilution of the flocculating cement particles in the HVFA mixture, the increase in HRWRA dosage accompanying this reduction in *w/cm* may not be as large as that which is needed in a non-fly ash concrete.

2) Type III cements can provide a significant boost to the early age strengths of HVFA concrete mixtures. The typical cost premium for switching to Type III cement is nominally only 10 % (cement cost basis).

3) In addition to their beneficial reduction of setting time delays, fine limestone additions can also significantly increase early-age and 28 d compressive strengths.

4) The above approaches to increasing early-age strengths are not mutually exclusive and any two or even all three can be employed in practice, depending on job requirements and materials' availability.

5) The maturity method that has been well-established for OPC concretes applies equally well to HVFA concrete mixtures. A unique feature observed for HVFA mortars is that, unlike their OPC counterparts, long-term strength is enhanced by higher temperature curing.

6) Particularly for HVFA concrete mixtures, the necessity of testing for strength compliance at 28 d should be considered carefully. As shown in Figure 10, HVFA mixtures continue to develop significant additional strength beyond 28 d, so that if compliance testing can be postponed to 56 d or even 91 d, it is more likely that these economical and environmentally friendly HVFA mixtures (with less cementitious material) can be employed in practice.

Transport Properties and Durability

With the exception of deicer scaling issues (to be discussed subsequently), generally, the transport properties and durability of properly designed and prepared HVFA concretes can be equal or superior to those of their corresponding 100 % OPC target mixtures. Often, lower replacements levels of fly ash (15 % to 25 %, for example) are used (or even actually mandated) to specifically address potential durability issues such as chloride diffusion, sulfate attack, or alkali-silica reaction by state DOT's and others [7,23].

Some representative electrical property results for the ternary blend HVFA concretes discussed previously (Table 3) are provided in Figure 14 and Figure 15 for RCPT and surface resistivity testing, respectively [65]. In addition to the nine mixtures detailed in Table 3, six additional ternary blend concrete mixtures were prepared. For the two 40 % replacement mixtures, w/cm was reduced by replacing water with sand, as detailed in Appendix C. For the 60 % replacement mixtures, mixture proportions were modified both by lowering w/cm (replacing water with the ternary blend of powders) and switching to a Type III cement, with two mixtures with differing w/cm being prepared for each fly ash. As can be observed in Figure 14, all of the ternary blends (with fine limestone) provide significant reductions in RCPT values, and corresponding increases in measured surface resistivities (Figure 15), both measured at 56 d. In seven cases, the HVFA concrete RCPT values are less than ½ of that measured for the reference 100 % OPC w/c=0.40 concrete. Likewise, in these same seven HVFA mixtures, the measured surface resistivities are more than double the value measured for the 100 % OPC mixture. These measured increases in resistivity are likely due to several contributing causes, including the increased reactions (reduced capillary porosity) in the presence of the fine limestone, the formation of more voluminous carboaluminate reactions products (less capillary porosity again) [60], the reduction in diffusivity of pozzolanic (lower Ca/Si ratio) calcium silicate hydrate gel (C-S-H) relative to that of C-S-H formed from conventional cement hydration [58], and the increased resistivity of the pore solution in the HVFA concrete due to cement dilution and increased sorption of alkali (Na^+ and K^+) ions by pozzolanic C-S-H and other reaction products formed in the presence of fly ash (and fine limestone) [72,73].

For mixtures containing fly ash, some of these same effects can contribute to their increased resistance to deleterious expansion caused by alkali-silica reactions (ASR). According to Shafaatian et al. [74], among the many potential contributors to this performance enhancement, there are four significant factors in mixtures containing fly ash: 1) increased alkali binding, 2) mass transport reduction, 3) increased tensile strength, and 4) a reduction in the aggregate dissolution rate (due to a localized reduction in pH near the reactive aggregate surfaces). The utilization of fly ash to mitigate ASR is well known, as exemplified by the ACI 232.2R-03 document presenting the utilization of Class F fly ash at a 20 % to 25 % replacement level (mass basis) as a general preventive measure to be employed when acceptable ASR performance of the aggregates being used in a particular concrete mixture cannot be guaranteed [23].

The performance of high-volume fly ash concretes in applications where de-icing chemicals are applied has been a subject of concern for many years [75]. Often, under laboratory testing conditions, HVFA concretes exhibit increased spalling and mass loss in comparison to

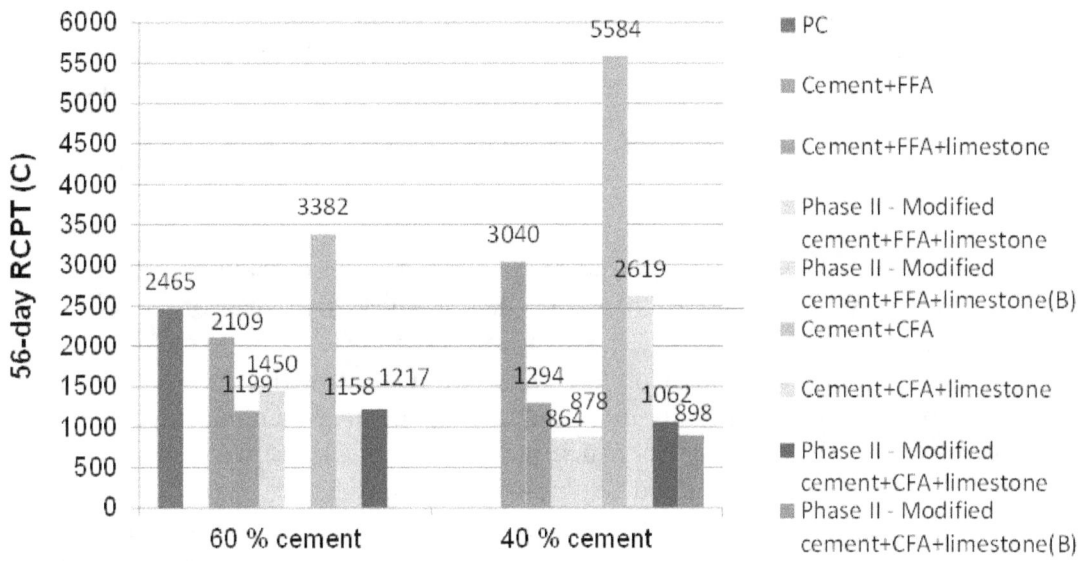

Figure 14. Measured cumulative charge passed (RCPT) at 56 d for the 15 concrete mixtures (taken from [65]). Coefficients of variation for three replicate specimens varied from 2.1 % to 19.1 % for the various mixtures.

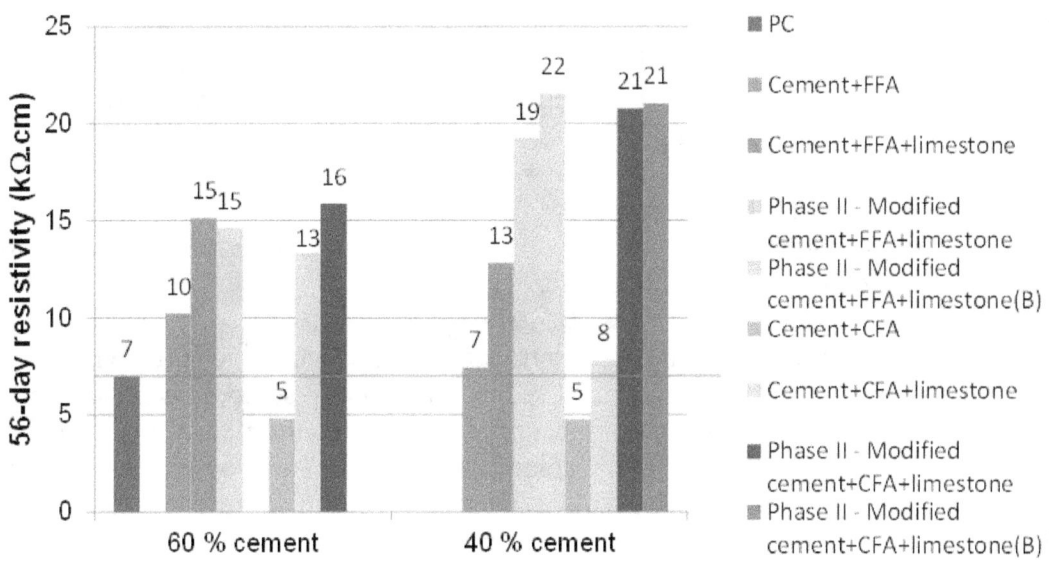

Figure 15. Measured surface resistivity at 56 d for 15 concrete mixtures (taken from [65]). Coefficients of variation for three replicate specimens varied from 0.6 % to 9.9 % for the various mixtures.

corresponding 100 % OPC concrete mixtures, typically with similar air contents in both types of mixtures. It should be noted that to achieve similar air contents, it is usually necessary to increase the dosage of the air-entraining agent (AEA) in the HVFA concrete mixture [76,77], due to increased absorption of the AEA by the porous fly ash particles, particularly those containing a substantial content of carbon. Ley [76] has developed a methodology for determining the required AEA dosage based on a single concrete mixture that should be equally applicable to HVFA mixtures. Recently, Van den Heede et al. [78] have developed an HVFA concrete mixture that meets the relevant European salt scaling criterion, with a minimum air content of 6 % to 7 %. On a per mass of binder basis, this formulation required 3.5 times the dosage of AEA typically used in a traditional control concrete. In such HVFA mixtures, air void stability is also a critical concern and the authors of the study recommended that for HVFA concrete mixtures, the air content be evaluated 60 min and 120 min after mixing, in addition to the conventional measurement conducted after 15 min.

Neuwald et al. [79] have hypothesized that the increased scaling behavior of HVFA concretes is due to the formation of a weak surface layer, one that often develops under laboratory conditions due to less than ideal curing, compounded with the inherently lower reactivity of most fly ashes in comparison to OPC. Poor curing practices produce a surface layer with enhanced porosity that is naturally more susceptible to subsequent scaling. Furthermore, in the presence of commonly employed chemical admixtures, such as AEAs and HRWRAs, the authors observed enhanced bleeding and separation of paste mixtures into two different density 'slurries,' with the top layer being less dense (and hence weaker). They concluded that these processes would therefore be more likely to occur in higher slump HVFA concrete mixtures. This separation would imply that this more porous top layer would 1) be able to absorb more of the deicer chemical solution and 2) would be weaker in resisting the (expansive) forces that occur upon freezing/thawing. Based on their observations, one could conclude that the scaling issues with HVFA concrete are not directly due to the presence of fly ash itself, but rather to issues of quality control and the lack of appropriate curing measures being applied in the field. Anecdotal evidence from numerous field exposures is that properly prepared, placed and cured HVFA concrete mixtures perform just as well as their OPC counterparts with respect to scaling in the presence of de-icing chemicals.

Relevance to Industry Practice:
1) In general, the transport properties and durability of properly cured HVFA concretes are superior to those of corresponding OPC mixtures.

2) Several factors contribute to the overall reduction in electrical conductivity (diffusivity) observed in HVFA concretes including a reduction in pore solution conductivity and the formation of pozzolanic C-S-H with an inherently lower conductivity than conventional C-S-H. In mixtures with fine limestone, additional contributions are provided by the enhanced degree of hydration and the formation of more voluminous carboaluminate reaction products, both of which decrease the capillary porosity of the ternary blend HVFA concrete mixture.

3) With respect to scaling in the presence of de-icing chemicals, inconsistencies remain between laboratory and field testing results and further research is needed to totally resolve these outstanding issues. The increased scaling of HVFA concretes in the laboratory seems to be often related to the formation of a weaker surface layer, due to less than ideal curing or the sensitivity of HVFA concretes with certain chemical admixtures to enhanced bleeding and settlement.

Mass Concrete Considerations

Some of the first applications of fly ash concretes were in the large dams constructed in the U.S between 1940 and 1970. In mass concrete, the slower (pozzolanic) reactions of the fly ash are a benefit in reducing the substantial early-age internal temperature rise that typically occurs within a mass OPC concrete. This reduced temperature rise will, in turn, reduce the propensity for: 1) early-age thermal cracking, 2) reduced in-place strengths due to higher temperature curing, and 3) the destabilization of ettringite (which can lead to problems with delayed ettringite formation, DEF, at a later age) [80]. This beneficial reduction in internal temperature is exemplified in the results provided in Figure 16 for HVFA mortars with and without IC [37]. At a fixed *w/cm* of 0.3 (likely below the value typically employed in mass concrete), all of the mixtures with fly ash exhibit a significantly lower and delayed temperature peak than that of the 100 % OPC control mortar. In comparison to the target *w/c*=0.42 control mortar, the 60 % replacement HVFA mortars exhibit about a 10 °C lower temperature peak that occurs after a 24 h delay with respect to the OPC concrete. The 40 % replacement HVFA mortars exhibit a similar temperature peak as the *w/c*=0.42 control OPC mortar, as a fairly reactive Class C fly ash was employed in the study, but the peak is delayed by approximately 12 h, mainly due to the retardation produced by this particular Class C fly ash. Similar performance is exhibited in concretes containing either a Class C or a Class F fly ash, along with a fine limestone, as shown in Figure 17. In terms of cracking susceptibility, both the temperature extremes that are achieved and the heating/cooling rates are critical parameters. In this regard, it is worth noting that concrete temperature and stress development can be modeled using a variety of freely available software packages, such as HIPERPAV III [81] and ConcreteWorks [82].

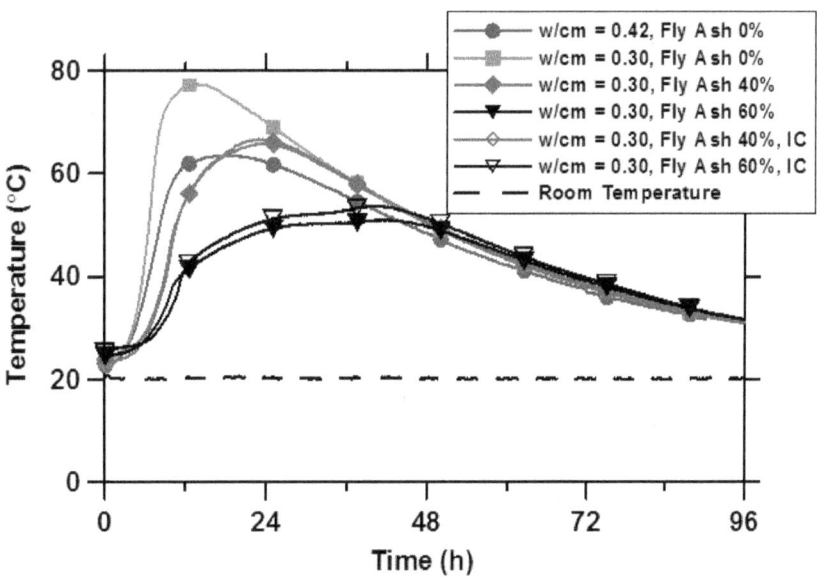

Figure 16. Semi-adiabatic temperature rise of mortars with various additions of fly ash, with and without IC [37].

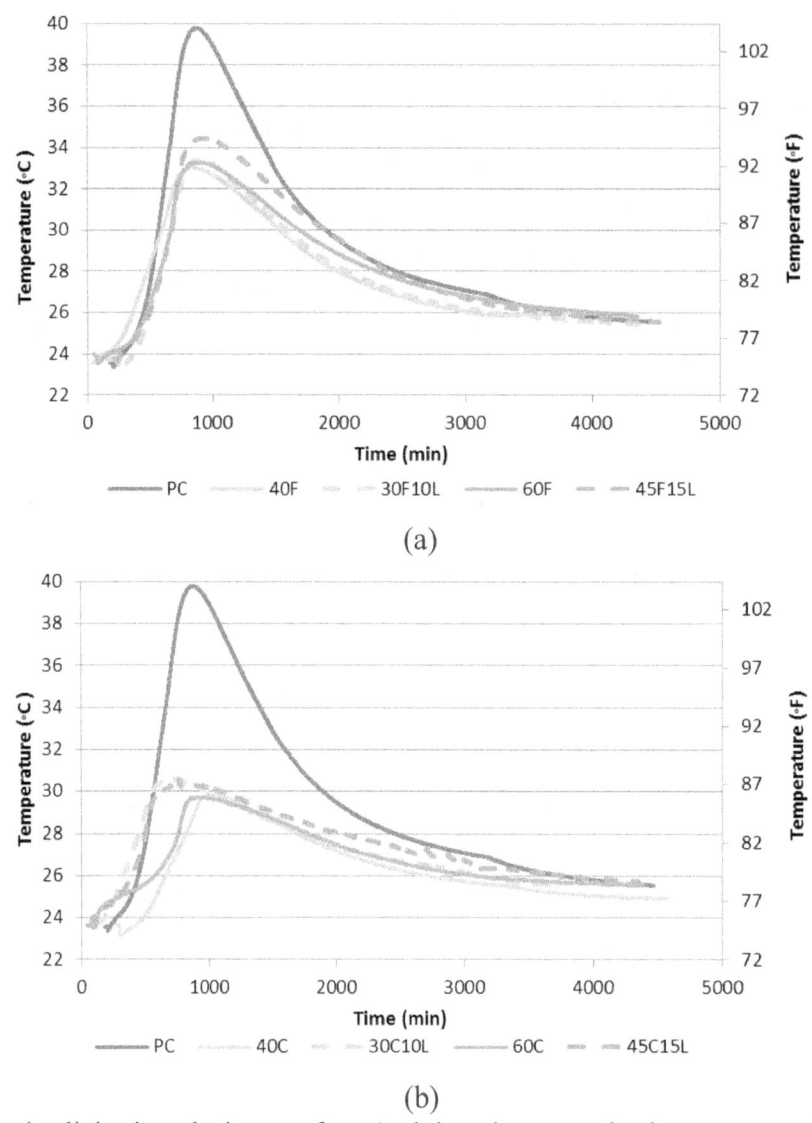

(a)

(b)

Figure 17. Semi-adiabatic calorimetry for: a) plain mixture and mixtures containing Class F fly ash and b) plain mixture and mixtures containing Class C fly ash [61]. Concrete mixture proportions can be found in Table 3.

In addition to reducing the rate of heat generation within a mass concrete element, the replacement of cement by fly ash will also change the thermophysical properties of the concrete mixture. This, along with the specific environmental boundary conditions, will in turn control how fast the generated heat can be dissipated from the mass concrete to its surroundings. While the specific heat capacity of HVFA mortars and concretes are similar to those of their OPC counterparts [83], both the density and the thermal conductivity of the HVFA composites are reduced due to the replacement of cement by the lower density fly ash. Because the thermal diffusivity that regulates how fast generated heat is dissipated is proportional to the ratio of these two properties, HVFA concretes could have a thermal diffusivity that is either higher or lower than that of a corresponding 100 % OPC concrete. In practice, measurements on a series of HVFA concretes with cement replacements levels varying from 0 % to 75 % on a mass basis and prepared with limestone aggregates have indicated that, for design purposes, a constant thermal

34

diffusivity value of about 1×10^{-6} m^2/s [83] can be employed. This is in contrast to a value of 1.5×10^{-6} m^2/s presented by Tatro [84] as a typical value for the thermal diffusivity of concretes with limestone aggregates.

With respect to energy efficiency, the insulating properties of a building material are typically characterized by its thermal conductivity (as opposed to diffusivity), commonly converted to an R-value for classification and comparison purposes. In this case, the thermal conductivity of HVFA mortars and concretes are typically 10 % to 50 % lower than those of their corresponding 100 % OPC counterparts [83], which should lead to savings in heating/cooling costs for residential and commercial structures constructed using HVFA concretes.

Relevance to Industry Practice:
1) Some of the earliest uses of fly ash in the U.S. were driven by the objective of reducing the temperature rise experienced in massive concrete structures (dams). HVFA concretes continue to be a viable option to limit the temperature rise experienced in mass concrete elements.

2) Computer software packages [81,82] are freely available for downloading over the Internet for estimating the maximum temperature rise that will be experienced in a specified mass concrete structure.

3) While the thermal conductivity of HVFA concrete will typically be lower than that of a corresponding OPC concrete, their specific heat capacities and thermal diffusivities will be fairly similar. The lower thermal conductivity of HVFA concrete will provide benefits in terms of increased insulating capability and reduced heating/cooling costs; this is another example of how HVFA concretes can make a positive contribution to the overall sustainability of our built environment.

Summary and Prospectus

The results presented in this guide have demonstrated that HVFA concretes can offer substantial performance benefits, in addition to their sustainability advantages in terms of reductions in cost, embodied energy, and CO_2 emissions. It must be recognized that HVFA concrete is a different material from conventional concrete and may therefore require new and/or different characterization, production, placement, and finishing procedures than have been conventionally employed by the construction industry. Published research data suggest that when proper attention is given to these concerns, HVFA concretes with equivalent or superior performance to that of the OPC concrete they are replacing can be achieved. While this report did not provide an extensive list of success stories using HVFA concretes, such a list is easily accessible in the fourth edition of the seminal reference on high-performance, high-volume fly ash concrete written by Malhotra and Mehta [1].

This report represents only a single snapshot in time with regards to the technology of HVFA concrete and its acceptance by the construction industry. As indicated by the 2012 NRMCA survey [2], fly ash usage in concrete has been continually increasing during the past 10 years and it is hoped that the technologies and guidance provided in this report can contribute to an acceleration of this growth in the near future.

Acknowledgements

The authors would like to thank Mr. Douglas Burke of Naval Facilities, Mr. Ryan Henkenseifken of U.S. Concrete, and Ms. Lisa Simpson of Heritage Concrete for providing valuable input to the section on current practices. They also wish to acknowledge useful discussions with Prof. Jason Weiss of Purdue University and Dr. Jussara Tanesi of SES Group and Associates, LLC.

References

1) V.M. Malhotra and P.K. Mehta, High-Performance, High-Volume Fly Ash Concrete for Building Durable and Sustainable Structures, 4th Edition, Supplementary Cementing Materials for Sustainable Development, Inc., Ottawa (2012).

2) K.H. Obla, C.L. Lobo, and H. Kim, The 2012 NRMCA supplementary cementitious materials use survey, Concr. Infocus, Summer 2012.

3) L. Simpson, personal communication, Feb. 16, 2012.

4) R. Henkenseifken, personal communication, March 7, 2012.

5) D. Cross, J. Stephens, and M. Berry, Sustainable construction contributions from the treasure state, Concr. Inter. **32** (5), 41-46 (2010).

6) D.F. Burke, Development of Concrete Mixtures with High-Volume Fly Ash Cement Replacement, 2012 International Concrete Sustainability Conference, National Ready Mixed Concrete Association (2012) 12 pp.

7) C. Rao, R.D. Stehly, and A. Ardani, Proportioning Fly Ash as Cementitious Material in Airfield Pavement Concrete Mixtures, Report IPRF-01-G-002-06-2, Innovative Pavement Research Foundation (2011).

8) ASTM International, ASTM C618-12a Standard Specification for Coal Fly Ash and Raw or Calcined Natural Pozzolan for Use in Concrete, West Conshohocken, PA, 2012.

9)) ASTM International, ASTM C311-11b Standard Test Methods for Sampling and Testing Fly Ash or Natural Pozzolans for Use in Portland-Cement Concrete, West Conshohocken, PA, 2011.

10) D.P. Bentz, A. Duran-Herrera, and D. Galvez-Moreno, Comparison of ASTM C311 strength activity index testing vs. testing based on constant volumetric proportions, J. ASTM Inter. **9** (1), 7 pp. (2012).

11) R.S. Winburn, S.L. Lerach, B.R. Jarabek, M.A. Wisdom, D.G. Grier, and G.J. McCarthy, Quantitative XRD analysis of coal combustion by-products by the Rietveld method. Testing with standard mixtures, Adv. X-ray Anal. **42**, 387-396 (2000).

12) D.P. Bentz, P.E. Stutzman, C.-J. Haecker, and S. Remond, SEM/X-ray imaging of cement-based materials, in Proceedings of the 7th Euroseminar on Microscopy Applied to Building Materials, Delft University of Technology, Delft (1999) 457-466.

13) B.G. Kutchko and A.G. Kim, Fly ash characterization by SEM-EDS, Fuel **85**, 2537-2544 (2006).

14) R.T. Chancey, P.E. Stutzman, M.C.G. Juenger, and D.W. Fowler, Comprehensive phase characterization of crystalline and amorphous phases of a Class F fly ash, Cem. Concr. Res. **40** (1), 146-156 (2010.)

15) K. Aughenbaugh, R.T. Chancey, P. Stutzman, M.C.G. Juenger, and D.W. Fowler, An examination of the reactivity of fly ash in cementitious pore solutions, Mater. Struct. 12 pp. (2012) doi: 10.1617/s11527-012-9939-6.

16) X. Feng, E.J. Garboczi, D.P. Bentz, P.E. Stutzman, and T.O. Mason, T.O., Estimation of the degree of hydration of blended cements by a scanning electron microscopy point-counting procedure, Cem. Concr. Res. **34** (10), 1787-1793 (2004).

17) M. Ben Haha, K., De Weerdt, and B. Lothenbach, Quantification of the degree of reaction of fly ash, Cem. Concr. Res. **40** (11), 1620-1629 (2010).

18) N. De Belie, G. Baert, and G. De Schutter, Modeling of microstructure of portland cement: Fly ash binders based on calorimetric and thermogravimetric experiments, in Proceedings of the 13th ICCC International Congress on the Chemistry of Cement, Madrid, Spain (2011) 1-7.

19) S. Ohsawa, K. Asaga, S. Goto, and M. Daimon, Quantitative determination of fly ash in hydrated fly ash-$CaSO_4$-$2H_2O$-$Ca(OH)_2$ system, Cem. Concr. Res. **15**, 357-366 (1985).

20) S. Li, D.M. Roy, and A. Kumar, Quantitative determination of pozzolans in hydrated systems of cement or $Ca(OH)_2$ with fly ash or silica fume, Cem. Concr. Res. **15**, 1079-1086 (1985).

21) L. Lam, Y. L. Wong, and C.S. Poon, Degree of hydration and gel/space ratio of high-volume fly ash/cement systems, Cem. Concr. Res. **30**, 747-756 (2000).

22) ACI Committee 211, ACI 211.1-91 (Reapproved 2009) Standard Practice for Selecting Proportions for Normal, Heavyweight, and Mass Concrete, American Concrete Institute, Farmington Hill, MI, 2009.

23) ACI Committee 232, ACI 232.2R-03 Use of Fly Ash in Concrete, American Concrete Institute, Farmington Hill, MI, 2003.

24) A. Schwartzentruber and C. Catherine, La méthode du mortier de béton équivalent (MBE) – Un nouvel outil d'aide á la formulation des bétons adjuvantés, Mater. Struct. **33**, 475-482 (2000).

25) E.P. Koehler and D.W. Fowler, ICAR Proportioning Procedure for Self-Consolidating Concrete, International Center for Aggregates Research, Austin, TX, 2007, 22 pp.

26) D.P. Bentz, C.F. Ferraris, I. De la Varga, M.A. Peltz, and J. Winpigler, Mixture proportioning options for improving high volume fly ash concretes, Inter. J. Pavement Res. Tech. **3** (5), 234-240 (2010).

27) ASTM International, ASTM C1679-09 Standard Practice for Measuring Hydration Kinetics of Hydraulic Cementitious Mixtures Using Isothermal Calorimetry, West Conshohocken, PA, 2009.

28) V.T. Cost and G. Knight, Use of thermal measurements to detect potential incompatibilities of common concrete materials, in Concrete Heat Development Monitoring Prediction and Management, ACI SP-241-4 (CD-ROM) (2007) 39-58.

29) M.D. Niemuth, L. Barcelo, and J. Weiss, Effect of fly ash on optimum sulfate levels measured using heat and strength at early ages, Adv. Civil Eng. Mater. **1** (1), 1-18 (2012) doi:10.1520/ACEM20120012.

30) D.P. Bentz, Powder additions to mitigate retardation in high volume fly ash mixtures, ACI Mater. J. **107** (5), 508-514 (2010).

31) D.P. Bentz, T. Sato, I. De la Varga, and J. Weiss, Fine limestone additions to regulate setting in high volume fly ash mixtures, Cem. Concr. Comp. **34** (1), 11-17 (2012).

32) L.R. Roberts and P.C. Taylor, Understanding cement-SCM-admixture interaction issues, Concr. Inter. **29** (1), 33-41 (2007).

33) J. Tanesi, A. Ardani, R. Meininger, and N. Nicolaescu, Evaluation of High-Volume Fly Ash (HVFA) Mixtures (Paste and Mortar Components) Using a Dynamic Shear Rheometer (DSR) and Isothermal Calorimeter, Report No. PB2012-112546, National Technical Information Service, Springfield, VA, 2012.

34) D.P. Bentz, Influence of Curing Conditions on Water Loss and Hydration in Cement Pastes with and without Fly Ash Substitution, NISTIR **6886**, U.S. Department of Commerce, July 2002.

35) T. Özturan and M.E. Baştopçu, Effect of curing on durability of fly ash concrete, in Durability of Concrete, Proceedings Sixth International Conference, ACI SP-212, American Concrete Institute, Farmington Hills, MI (2003) 353-368.

36) D.P. Bentz and W.J. Weiss, Internal Curing: A 2010 State-of-the-Art Review, NISTIR **7765**, U.S. Department of Commerce, February 2011.

37) I. De la Varga, J. Castro, D. Bentz, and J. Weiss, Application of internal curing for mixtures containing high volumes of fly ash, Cem. Concr. Comp. **34** (9), 1001-1008 (2012).

38) B. Mather, Self curing concrete, why not?, Concr. Inter. **23** (1), 46-47 (2001).

39) C.F. Ferraris and F. de Larrard, Testing and Modelling of Fresh Concrete Rheology, NISTIR **6094**, U.S. Department of Commerce, February 1998. Available at: http://ciks.cbt.nist.gov/~garbocz/rheologyNISTIR/

40) N. Roussel and P. Coussot, Fifty-cent rheometer" for yield stress measurements: From slump to spreading flow, J. Rheol. **49**, 705 (2005) doi: 10.1122/1.1879041.

41) S. Chidiac, F. Habibbeigi, and D. Chan, Slump and slump flow for characterizing yield value of fresh concrete, ACI Mater. J. **103**, 413-418 (2006).

42) C.F. Ferraris and L. Brower, eds., Comparison of Concrete Rheometers: International Tests at LCPC (Nantes, France) in October 2000, NISTIR **6819**, U.S. Department of Commerce, September 2001. Available at: http://fire.nist.gov/bfrlpubs/build01/PDF/b01074.pdf

43) C.F. Ferraris and L. Brower, eds., Comparison of Concrete Rheometers: International Tests at MB (Cleveland OH, USA) in May 2003, NISTIR **7154**, U.S. Department of Commerce, September 2004. Available at: http://ciks.cbt.nist.gov/~ferraris/PDF/DraftRheo2003V11.4.pdf

44) C.F. Ferraris, K. Obla, and R. Hill, The influence of mineral admixtures on the rheology of cement paste and concrete, Cem. Concr. Res. **31** (2), 245-255 (2001).

45) K.H. Khayat, A. Yahia, and M. Sayed, Effect of supplementary cementitious materials on rheological properties, bleeding, and strength of structural grouts, ACI Mater. J. **105**, 585-593 (2008).

46) V.G. Jimenez-Quero, F.M. Leon-Martinez, P. Montes-Garcia, C. Gaona-Tiburcio, and J.G. Chacon-Nava, Influence of sugar-cane ash and fly ash on the rheological behavior of cement pastes and mortars, Constr. Building Mater. **40**, 691-701 (2013).

47) S.H. Lee, H.J. Kim, E. Sakai, and M. Daimon, Effect of particle size distribution of fly ash–cement system on the fluidity of cement pastes, Cem. Concr. Res. **33**, 763-768 (2003).

48) A.K.H. Kwan and Y. Li, Effects of fly ash microsphere on rheology, adhesiveness and strength of mortar, Constr. Building Mater. **42**, 137-145 (2013).

49) P. Termkhajornkit and T. Nawa, The fluidity of fly ash-cement paste containing naphthalene sulfonate superplasticizer, Cem. Concr. Res. **34**, 1017-1024 (2004).

50) D.P. Bentz, C.F. Ferraris, M.A. Galler, A.S. Hansen, and J.M. Guynn, Influence of particle size distributions on yield stress and viscosity of cement-fly ash pastes, Cem. Concr. Res. **42** (2), 404-409 (2012).

51) N.A. Libre, R. Khoshnazar, and M. Shekarchi, Relationship between fluidity and stability of self-consolidating mortar incorporating chemical and mineral admixtures, Constr. Building Mater. **24** 1262-1271 (2010).

52) V.T. Cost, Concrete sustainability versus constructability – Closing the gap, in Proceedings of the 2011 International Concrete Sustainability Conference, Boston, MA (2011) available at: http://www.nrmcaevents.org/?nav=display&file=189.

53) K.P. Keith and A.K. Schlinder, Phase II – Task 3 Setting and Temperature Development, Final Report for Project BAA No. DTFH61-08-R-0034 Greatly Increased Use of Fly Ash in Hydraulic Cement Concrete (HCC) for Pavement Layers and Transportation Structures (2012).

54) D.P. Bentz and C.F. Ferraris, Rheology and setting of high volume fly ash mixtures, Cem. Concr. Comp. **32** (4), 265-270 (2010).

55) T.J. Looney, M. Arezoumandi, J.S. Volz, and J.J. Myers, An experimental study on bond strength of reinforcing steel in high-volume fly-ash concrete, Adv. Civil Eng. Mater. **1** (1), 1-17 (2012) doi:10.1520/ACEM20120026.

56) V.T. Cost and P. Bohme, Synergies of portland-limestone cements and their potential for concrete performance enhancement, in Proceedings of the 2012 International Concrete Sustainability Conference, Seattle, WA (2012).

57) L. Gurney, D.P. Bentz, T. Sato, and W.J. Weiss, Using limestone to reduce set retardation in high volume fly ash mixtures: Improving constructability for sustainability, Trans. Res. Rec.: J. Trans. Res. Board **2290**, 139-146 (2012).

58) K. De Weerdt, K.O. Kjellsen, E. Sellevold, and H. Justnes, Synergy between fly ash and limestone powder in ternary cements, Cem. Concr. Comp. **33** (1), 30-38 (2011).

59) K. De Weerdt, M. Ben Haha, G. Le Saout, K.O. Kjellsen, H. Justnes, and B. Lothenbach, Hydration mechanisms of ternary portland cements containing limestone powder and fly ash, Cem. Concr. Res. **41** (3), 279-291 (2011).

60) P. Mounanga, M.I.A. Khokhar, R. El Hachem, and A. Loukili, Improvement of the early-age reactivity of fly ash and blast furnace slag cementitious systems using limestone filler, Mater. Struct. **44**, 437-453 (2011).

61) J. Tanesi, D.P. Bentz, and A. Ardani, Enhancing high volume fly ash concretes using fine limestone powder, in Green Cements, ACI SP - , Minneapolis, MN (2013).

62) J. Moon, J.E. Oh, M. Balonis, F.P. Glasser, S.M. Clark, and P.J.M. Monteiro, High pressure study of low compressibility tetracalcium aluminum carbonate hydrates $3CaO \cdot Al_2O_3 \cdot CaCO_3 \cdot 11H_2O$, Cem. Concr. Res. **42**, 105-110 (2012).

63) D.A. Abrams, Design of Concrete Mixtures, Bulletin 1, Structural Materials Research Laboratory, Lewis Institute, Chicago, 1918, 20 pp.

64) R. Feret, Sur la compacité des mortiers hydraulique, Ann. Ponts Chaussées **7** (4), 5-164 (1892).

65) D.P. Bentz, J. Tanesi, and A. Ardani, A., Ternary blends for controlling cost and carbon content: High-volume fly ash mixtures can be enhanced with additions of limestone powder, Concr. Inter. **35** (8), 51-59 (2013).

66) K. Obla, S. Upadhyaya, D. Goulias, A. Schindler, and N.J. Carino, New Technology-Based Approach to Advance Higher Volume Fly Ash Concrete with Acceptable Performance, Final Report, National Ready-Mixed Concrete Association, August 2008.

67) N.J. Carino, The maturity method: Theory and application, Cem. Concr. Agg. **6** (2), 61-73 (1984).

68) ASTM International, ASTM C1074-11 Standard Practice for Estimating Concrete Strength by the Maturity Method, West Conshohocken, PA, 2011.

69) D.P. Bentz, T. Barrett, I. De la Varga, and W.J. Weiss, Relating compressive strength to heat release in mortars, Adv. Civil Eng. Mater. **1** (1), 1-14 (2012) doi:10.1520/ACEM20120002.

70) T.J. Barrett and W.J. Weiss Performance of More Sustainable Cements that Include Interground Limestone Additions of up to 15 %, Joint Transportation Research Program, 2012.

71) D.P. Bentz, O.M. Jensen, A.M. Coats, and F.P. Glasser, Influence of silica fume on diffusivity in cement-based materials. I. Experimental and computer modeling studies on cement pastes, Cem. Concr. Res. **30** (7), 1121-1129 (2000).

72) H.F.W. Taylor, A method for predicting alkali ion concentrations in cement pore solutions, Adv. Cem. Res. **1** (1), 5-16 (1987).

73) E. Schafer and B. Meng, Influence of cement and additions on the quantity of alkalis available for an alkali-silica reaction, Beton, 577-584 (2001).

74) S.M.H. Shafaatian, A. Akhavan, H. Maraghechi, and F. Rajabipour, How does fly ash mitigate alkali-silica reaction (ASR) in accelerated mortar bar test (ASTM C1567), Cem. Concr. Comp. **37**, 143-157 (2013).

75) W.F. Perenchio and P. Klieger, Further Laboratory Studies of Portland-Pozzolan Cements, RD041.01T, Portland Cement Association, Skokie, IL, 1976.

76) M.T. Ley, Determining the air-entraining admixture dosage response for concrete with a single concrete mixture, J. ASTM Inter. **7** (2), 11 pp. (2010).

77) T. Ley, N.J. Harris, K.J. Folliard, and K.C. Hover, Investigation of required air entraining admixture dosage in fly ash concrete, ACI Mater. J. **105** (5), 494-498 (2008).

78) P. Van den Heede, J. Furniere, and N. De Belie, Influence of air entraining agents on deicing salt scaling resistance and transport properties of high-volume fly ash concrete, Cem. Concr. Comp. **37**, 293-303 (2013).

79) A. Neuwald, A. Krishnan, J. Weiss, J. Olek, and T. Nantung, Concrete curing and its relationship to measured scaling in concrete containing fly ash, in Transportation Research Board 82nd Annual Meeting Compendium of papers (CD-ROM), Washington D.C., January 12-16, 2003.

80) ACI Committee 207, ACI 207.1R-05 Guide to Mass Concrete, American Concrete Institute, Farmington Hills, MI, 2005.

81) http://www.hiperpav.com [accessed 02/21/13].

82) http://www.texasconcreteworks.com [accessed 02/21/13].

83) D.P. Bentz, M.A. Peltz, A. Durán-Herrera, P. Valdez, and C. Juárez, Thermal properties of high-volume fly ash mortars and concretes, J. Building Phys. **34** (3), 263-275 (2011).

84) S.B. Tatro, Thermal Properties, In: J.F. Lamond and J.H. Pielert (eds.) ASTM STP 169D-Significance of Tests and Properties of Concrete & Concrete Making Materials, Chapter 22, ASTM International, West Conshohocken, PA (2006) 226-237.

Appendix A. Flowchart for Proportioning HVFA Concrete Mixtures

Mixture Proportioning Procedure

1. Identify the compressive strength requirement(s): cylinder strength *vs.* age
2. Identify the components and proportions of a trial mixture (for estimating an initial trial *w/cm* see Appendix D)
3. Cast pastes and mortars:
   a. Use calorimetry to achieve optimal sulfate content and ensure compatibility (ASTM C1679 and ASTM C1702)
   b. Determine the initial and final setting times (ASTM C191 or ASTM C403)
   c. Use rheometry to preliminarily evaluate workability
4. Prepare concrete and cast cylinders, prisms, and rings
   a. Measure unit weight, air content, slump (and workability)
   b. Sieve mortars to verify setting times (ASTM C403) and measure autogenous shrinkage (ASTM C1698)
   c. Determine strength *vs.* age (ASTM C39)
   d. Shrinkage test (ASTM C157) on concrete prisms to evaluate drying shrinkage
   e. Ring test (ASTM C1581) on concrete to evaluate cracking propensity
   f. Test for any other performance specifications
5. Modify mixture to meet performance specifications
   a. Long-Term Strength: modify *w/cm* and/or replace part of fly ash with fine limestone
   b. Early-Age Strength: switch to a Type III cement, reduce *w/cm*, and/or use a chemical accelerator
   c. Setting Times: replace part of fly ash with fine limestone powder
   d. Shrinkage: employ internal curing (ASTM C1608 and ASTM C1791) or shrinkage-reducing admixture (SRA)
   e. Rheology: modify HRWA dosage

This process is shown schematically below as a flow table (adapted from reference 26).

45

High-Volume Fly Ash Concrete Mixture Proportioning

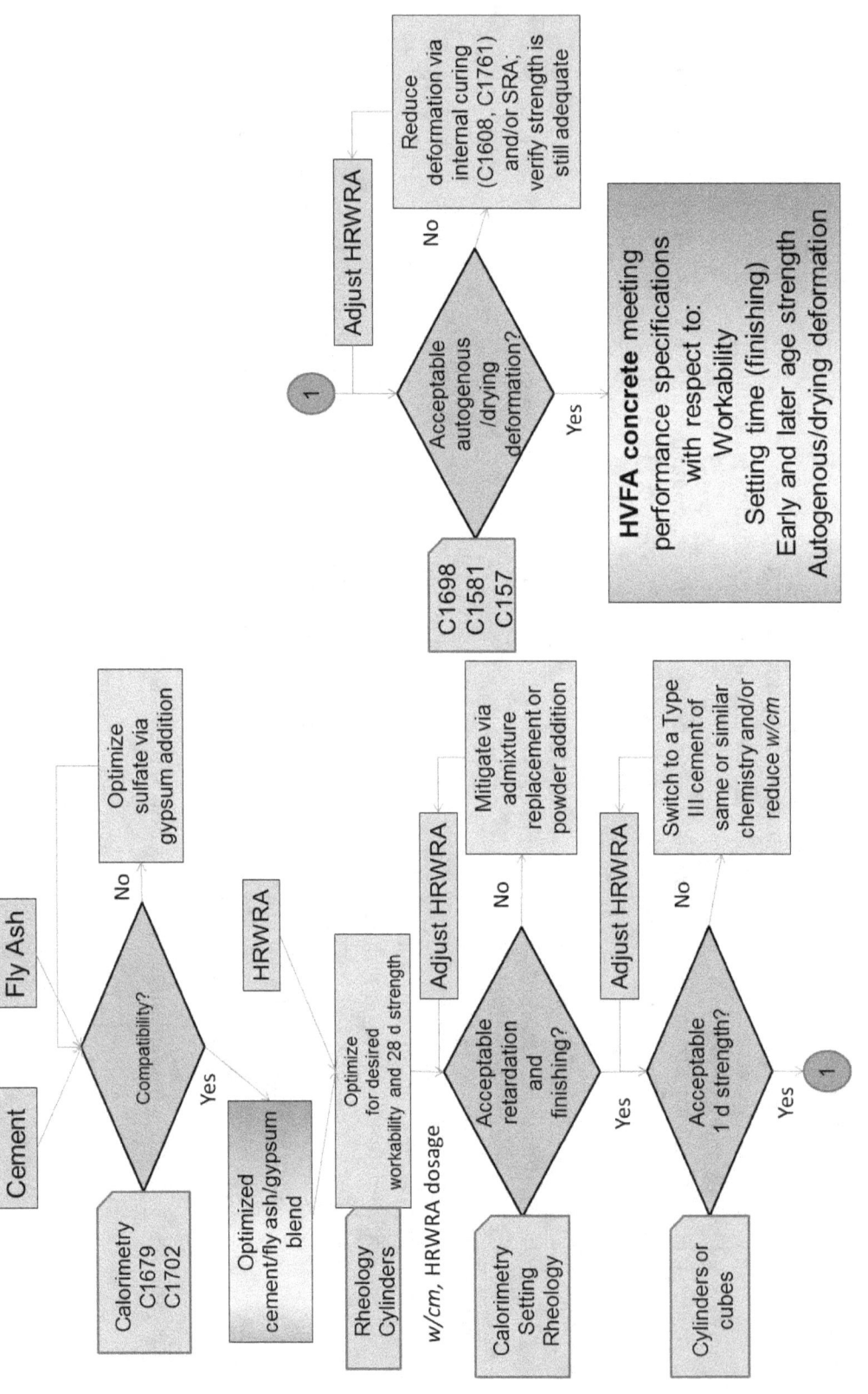

Appendix B. Steps in Proportioning HVFA Concrete Mixtures with Acceptable (Equivalent) Early-Age Performance

Performance characteristics to achieve (meet specified targets):
- Slump
- Air content
- Setting time (finishing, etc.)
- Early-age strength (1 d, 3 d, etc.)
- 28 d strength
- 28 d durability (rapid chloride penetrability (RCPT) or other)
- Drying shrinkage at 28 d or 56 d

Preliminary steps:
1) Characterize all powders (cement, fly ash, limestone, others) with respect to their specific gravity (ASTM C188 [B1] or other)[3]
2) Assess cement/fly ash/admixture compatibility using calorimetry (ASTM C1679 standard practice for isothermal calorimetry [B2] or the semi-adiabatic equivalent that is currently under draft by ASTM); find a set of compatible materials including high range water reducer (HRWR) for expected proportions of cement and fly ash (and other supplementary cementitious materials, SCM), a trial water-to-cementitious materials ratio (w/cm), and expected HRWR dosage

Mixture proportioning:
1) Use a volumetric-based proportioning method such as the aggregate suspension proportioning method developed by Koehler [B3], with recommended air entrainer addition (from chemical admixture manufacturer) to achieve targeted air content,
2) Specify desired volumetric replacement(s) of cement by other powders (fly ash, limestone, etc.),
3) Select a trial w/cm ratio and determine volume fractions of water and all cementitious (powder) components,
4) Prepare trial mixture and measure performance characteristics,
5) Adjust mixture proportions to obtain desired characteristics,
   a) if setting time is too long, it can be decreased by replacing a portion of the fly ash with fine limestone powder (optimally about 1 μm median diameter powder); usually in the 5 % to 15 % of total powder volume range
   b) if early age strength is too low options include
       i) switch to a Type III cement and first re-evaluate existing mixture with Type III,
       ii) lower w/cm (or water-to-cement ratio, w/c) based on a plot of strength vs. heat release per mL of water (see Appendix C for the detailed derivations); by

---

[3] Helium pycnometry can also be employed to measure particle (powder) density. While differences between helium pycnometry and ASTM C188 values for cements are generally on the order of 2 % or less, for more porous fly ashes, these differences are typically on the order of 10 % (with the C188 test method consistently giving a lower density than the helium pycnometry, due to the slow absorption of the C188 immersion fluid by the fine pores/voids within the fly ash particles). It is likely the C188 test method value that is of relevance with respect to volumetric-based mixture proportioning.

decreasing water and increasing cement (cementitious binder) volume fractions, the necessary w/c to achieve a projected desired cumulative heat release can be computed as a change in volume fractions of cement and water as detailed in Appendix C

6) Prepare revised mixture proportions and adjust slump to meet specified target via increase/decrease in HRWR(s) (be sure that all ingredients remain compatible at the new HRWR dosage(s)). With these adjustments, 28 d strength and RCPT will seldom have a problem in meeting specified values (however, if they do, further reductions in w/cm may be warranted).

7) If measured drying shrinkage is unacceptable, options include incorporating a shrinkage-reducing admixture (be sure to re-evaluate compatibility of all ingredients together), providing internal curing via the addition of pre-wetted lightweight fine aggregates, or some combination of these two.

References:

B1) ASTM C188-09 Standard Test Method for Density of Hydraulic Cement, ASTM International, West Conshohocken, PA, 2009.

B2) ASTM C1679-09 Standard Practice for Measuring Hydration Kinetics of Hydraulic Cementitious Mixtures Using Isothermal Calorimetry, ASTM International, West Conshohocken, PA, 2009.

B3) E.P. Koehler and D.W. Fowler, ICAR Proportioning Procedure for Self-Consolidating Concrete, International Center for Aggregates Research, Austin, TX, 2007, 22 pp.

Appendix C. Adjusting Mixture Proportions (*w/cm*) Based on Heat Release Data and Desired Early-Age Strength

Assumptions: In the following derivation, it is assumed that heat release at early ages (1 d or 3 d) is controlled by cement hydration with only minor contributions coming from fly ash (and/or limestone) reactions. If these materials are contributing to early-age strength, the presented procedure should be conservative and the obtained strengths should further exceed the desired levels. It is further assumed that a linear relationship has been established between cumulative heat release on a per mL mixing water basis and measured compressive strengths for the concrete of interest [C1].

Based on the measured and desired (projected) compressive strengths, the user determines the value of HR^p, the projected heat release needed to provide the projected compressive strength (from the established linear relationship; see Figure 12 and Figure 13 for examples). Heat release per mL water is proportional to cement content and inversely proportional to water content (volume fraction).

Thus, the derivation begins from:

$$HR^p = HR^m \frac{V_w^m}{V_w^p} \frac{V_c^p}{V_c^m} \tag{C-1}$$

where HR is the cumulative heat release in J/mL water, V is volume fraction, the subscripts w and c refer to water and cement, respectively, and the superscripts p and m refer to projected (desired) and measured, respectively.

To achieve the desired strength, the simplest approach is to replace water by cement (assuming that more strength is needed; if the converse is the case, one could replace cement by water or consider replacing cement by fly ash and/or limestone in which case equation C-1 would not contain the water volume fraction ratio term).

But, if replacing cement by water on a volumetric basis, one has

$$V_c^p = V_c^m + \Delta V \tag{C-2}$$
$$V_w^p = V_w^m - \Delta V \tag{C-3}$$

where ΔV is the computed volume fraction change for which a solution is desired.

Equation C-1 then becomes

$$HR^p = HR^m \frac{V_w^m}{V_w^m - \Delta V} \frac{V_c^m + \Delta V}{V_c^m} \tag{C-4}$$

Equation C-4 can be solved for ΔV, as

$$\Delta V = \frac{V_c^m (\frac{HR^p}{HR^m} - 1)}{1 + \frac{HR^p V_c^m}{HR^m V_w^m}} \tag{C-5}$$

50

Then, the new volume fractions of cement and water can be obtained from equations C-2 and C-3. Note that this replacement of water with cement will change the volumetric ratio of cement: fly ash: limestone in a ternary blend, etc.

Alternatively, if one desires to keep the volumetric ratios of cement: fly ash: limestone constant as the concrete mixture is redesigned to increase strength (perhaps a more sustainable approach), equations C-2 and C-3 would become:

$$V_c^p = V_c^m + \gamma \Delta V \tag{C-2a}$$
$$V_w^p = V_w^m - \Delta V \tag{C-3a}$$

where γ is the volume fraction of cement per total volume of solids (cement, fly ash, and/or limestone in the current mixture proportions). Then,

$$HR^p = HR^m \frac{V_w^m}{V_w^m - \Delta V} \frac{V_c^m + \gamma \Delta V}{V_c^m} \tag{C-4a}$$

and

$$\Delta V = \frac{V_c^m (\frac{HR^p}{HR^m} - 1)}{\gamma + \frac{HR^p V_c^m}{HR^m V_w^m}} \tag{C-5a}$$

Fly ash and limestone would also be increased to maintain the constant volumetric ratio of cement: fly ash: limestone as,

$$V_{FA}^p = V_{FA}^m + \alpha \Delta V \tag{C-6a}$$
$$V_{LP}^p = V_{LP}^m + \beta \Delta V \tag{C-7a}$$

where the subscripts FA and LP refer to fly ash and limestone powder, respectively, and α and β refer to the volume fractions of fly ash and limestone in the total powder, respectively. By definition, $\alpha + \beta + \gamma = 1$.

A third option to consider is the replacement of water by fine aggregate, so as not to have to increase the cement content of the mixture at all [C2]. In this case, one has:

$$V_c^p = V_c^m \tag{C-2b}$$
$$V_w^p = V_w^m - \Delta V \tag{C-3b}$$

$$HR^p = HR^m \frac{V_w^m}{V_w^m - \Delta V} \tag{C-4b}$$

$$\Delta V = V_w^m \left(\frac{HR^p}{HR^m} - 1 \right) \tag{C-5b}$$

References:

C1) D.P. Bentz, T. Barrett, I. De la Varga, and W.J. Weiss, Relating compressive strength to heat release in mortars, Adv. Civil Eng. Mater. **1** (1), 1-14 (2012) doi:10.1520/ACEM20120002.

C2) D. Bentz, J. Tanesi, and A. Ardani, Ternary blends for controlling cost and carbon content: High-volume fly ash mixtures can be enhanced with additions of limestone powder, Conc. Inter., **35** (8), 51-59 (2013).

Appendix D. Choosing an Initial Trial *w/cm* Based on Strength-Heat Release Relationships

Often, a ready-mix producer will have an intuitive notion for the initial *w/cm* to use in a trial HVFA concrete mixture in pursuit of a given set of strength criteria. Generally, this intuition will be based on experience with local materials and the performance of past OPC and lower fly ash content blended cement concrete mixtures. When such intuition (experience) is lacking, one approach to choosing an initial *w/cm* for trial mixtures is based on the strength-heat release relationships mentioned previously. This approach assumes that to an age of 28 d, the primary contributor to the strength development in an HVFA concrete mixture is the reaction of the portland cement component, with fly ash and limestone powder potentially contributing by accelerating the cement hydration and/or participating in the hydration and pozzolanic reactions.

Some example data sets illustrating recently measured strength-heat release relationships are provided in Figure 18. The data points represent results for concretes prepared at the Turner-Fairbanks Highway Research Center (TFHRC) in a recent Federal Highway Administration/National Institute of Standards and Technology (FHWA/NIST) collaborative study [65], where 14 different HVFA mixtures and two OPC (control) mixtures were evaluated for compressive strength and cumulative heat release, the latter always being measured on mortar specimens (prepared directly or sieved from the concrete). For these mixtures, both the coarse and fine aggregates were siliceous. In Figure 18, the line indicated as 'Purdue' data [70] represents the best fit [69] relationship for compressive strengths measured on OPC concretes with and without limestone powder replacement for cement, with heat release measured on equivalent mortars. In this case, the concrete mixtures were all prepared with a limestone coarse aggregate and silica sand. For both concrete data sets, no air-entraining agents were employed, but an HRWRA (different for the two sets of concretes) was used to obtain slumps on the order of 50 mm to 100 mm. To first order, the results presented in Figure 18 suggest that at equivalent heat release values, the compressive strength of concrete with the limestone coarse aggregate is about 1000 psi (6.895 MPa) higher than that of one prepared with the siliceous aggregates, in general agreement with results reported previously and usually attributed to better bonding between the hydrating cement paste and the limestone aggregate [D1] (although any elastic moduli differences between the two types of aggregates could also contribute to measured strength differences); of course, for the data in Figure 18, the concretes were also prepared with different cements, water sources, etc.

Using Figure 18, or its equivalent for local materials when available, an approach to estimating an initial *w/cm* for trial mixtures can be formulated as follows. An equation must be available relating strength to heat release, along with some estimate of the heat release (J/g cement) vs. curing age characteristics of the cement being employed in the mixtures. For example, for the TFHRC concrete data in Figure 18, the best fit linear relationship is given by:

$$\text{strength(psi)} = (-2046.1) + 10.408 * \text{heat(J/mL water)} \tag{D-1}.$$

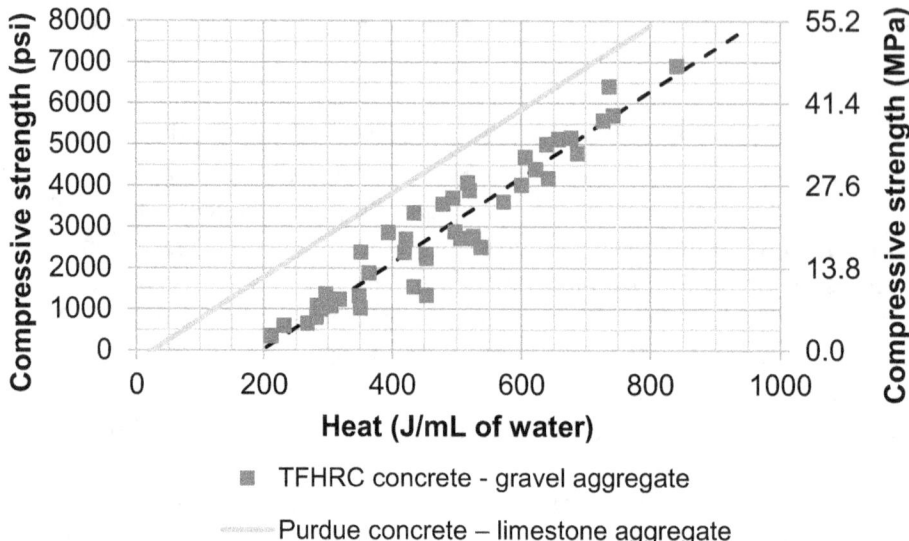

Figure 18. Measured concrete compressive strength vs. measured cumulative heat release (per unit volume of water) used to establish a strength-heat release linear relationship (dashed line, R^2=0.91) for choosing mixture proportions [65]. Purdue data is taken from [69].

Alternately, for the concretes with a limestone aggregate (Purdue data in Figure 18), the linear relationship is shown as:

$$\text{strength (psi)} = (-254.28) + 10.167 * \text{heat(J/mL water)} \tag{D-2}.$$

In either of these equations, the heat(J/mL water) term can be replaced by:

$$\text{heat(J/mL water)} = \text{heat(J/g cement)} * (1 + X_{FA}) * (1 + X_{LP}) / (w/c) \tag{D-3}$$

where heat(J/g cement) is the measured, known, or hypothesized heat release per unit mass of cement, X represents a strength enhancement factor for the fly ash (FA) and limestone powder (LP, when present) components of a ternary blend and w/c represents the mass-based ratio of water to **cement**, as usual. Here, we are approximating the density of water to be 1000 kg/m^3 (or 1 g/mL). The potential exists for estimating the required strength enhancement factor for the fly ash from the results of ASTM C311 strength activity index testing, particularly if this testing were to be converted from mass-based replacement and constant flow conditions to volume-based replacements and constant volume proportions as discussed in the Characterization section of this report [10]. A calibration of this approach to the TFHRC data in Figure 18 has produced the strength enhancement factors provided in Table 4 for the specific materials employed in that study [65]. These factors could vary significantly depending on the characteristics of the specific fly ash and limestone powder being utilized and therefore, the values in Table 4 should only be viewed as generic approximations to be employed in the absence of more definitive information.

For each age for which a required compressive strength is specified, an equation similar to D-1 or D-2 can be solved for the necessary heat release value and then equation D-3 solved for the necessary value of w/c. Once these w/c values are obtained for the different required ages, the minimum value from the set could be selected as the initial trial *w/c* for the mixture. This would then be converted to volumetric fractions of water, cement, fly ash, and limestone powder (when present), using the desired volumetric proportions of cement:fly ash:limestone. These volumetric proportions could also be used to compute the *w/cm* and water-to-powder ratio (*w/p*) of the trial mixture, if desired. However, to solve equation D-3 for *w/c*, the heat release of the cement being used as a function of curing age must be known or estimated. In the absence of measured data for the cement of interest, recommended ranges of values for Type I/II and Type III cements (based on cement pastes of these types with *w/c*=0.4 measured via isothermal calorimetry at NIST and FHWA) are provided in Table 5. If the trial mixture water content is too low to produce a workable concrete (even with the use of HRWRA), options would include switching to a Type III cement or incorporating fine limestone powder into the mixture, if these strategies were not already being employed in the proposed mixture proportions.

Table 4. Strength Enhancement Factors vs. Age as Estimated from Data in [65].

Age (d)	Class F fly ash (X_{FFA})	Class C fly ash (X_{CFA})	Limestone powder (X_{LP})
1	0.0	0.12	0.12
3	0.0	0.12	0.12
7	0.0	0.12	0.12
28	0.08	0.20	0.12

Table 5. Recommended Heat Release Ranges vs. Age For Various ASTM C150 Cement Types.

Age (d)	Type I/II cement	Type III cement
1	190 J/g to 220 J/g	250 J/g to 285 J/g
3	240 J/g to 300 J/g	275 J/g to 350 J/g
7	275 J/g to 330 J/g	305 J/g to 365 J/g
28	350 J/g	375 J/g

An example of employing this approach with the FHWA/NIST concrete data is provided in Figure 19. It must be kept in mind that the strength enhancement factors were calibrated using the results for the mixtures prepared with the Type I/II cement from this same data set. Still, the predicted strength values are quite close to the measured values, the former generally falling within 500 psi (3.45 MPa) of the latter. The worst predictions are obtained for the lower *w/cm* (≤ 0.4) mixtures using the Class C fly ash, where the measured strengths are significantly higher than those predicted by this approach. This is likely due to alkali activation of this fly ash that would be increased as *w/cm* is lowered and that is not accounted for in the current approach. However, because the measured strengths are higher than the predicted values, this approach to selecting an initial *w/c* (*w/cm*) for a trial mixture would still be a conservative one.

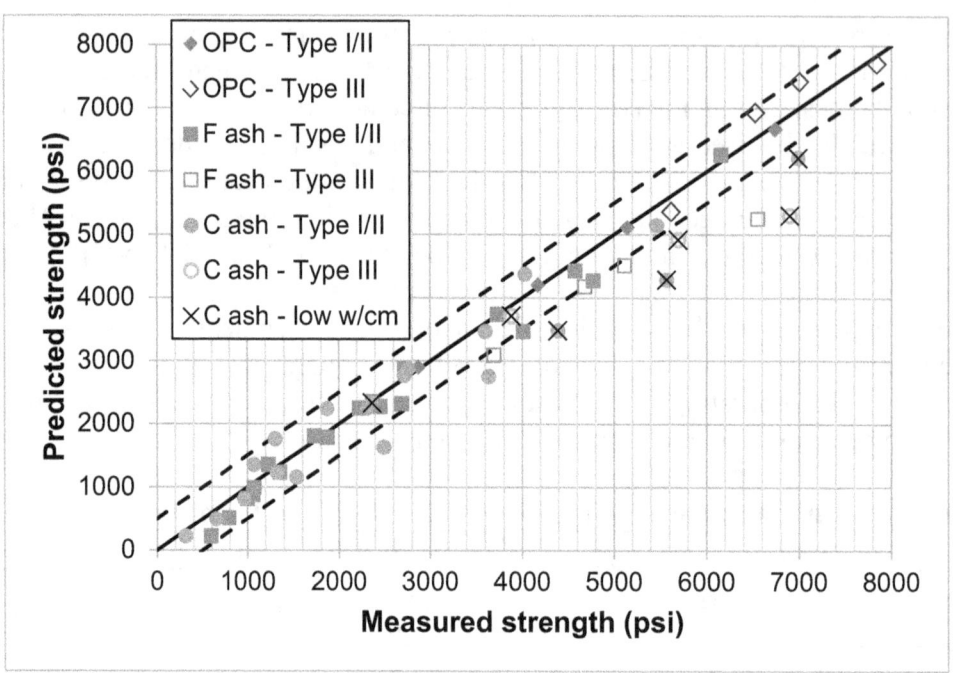

Figure 19. Measured vs. predicted strength for concrete mixtures from the FHWA/NIST study [65], based on the outlined approach. 1000 psi = 6.895 MPa. The solid line indicates a one-to-one relationship while the dashed lines represent ± 500 psi from this line of equality.

References:

D1) C.W. French and A. Mokhtarzadeh, High strength concrete: Effects of materials, curing and test procedures on short-term compressive strength, PCI J. **38** (3), 76-87 (1993).